THE MEANING OF FOSSILS

Episodes in the History of Palaeontology

Second Edition

MARTIN J. S. RUDWICK

The University of Chicago Press
Chicago and London

The University of Chicago Press, Chicago 60637
The University of Chicago Press, Ltd., London

 First edition 1972
University of Chicago Press edition 1985
Printed in the United States of America

94 93 92 91 90 89 88 87 86 85   5 4 3 2 1

Library of Congress Cataloging in Publication Data

Rudwick, M. J. S.
The meaning of fossils.

Reprint. Originally published: New York : Science History Publications. 2d. ed. 1976.
Bibliography: p.
Includes index.
1. Paleontology—History. I. Title.
QE705.A1R8 1985   560′.903   84-28080
ISBN 0-226-73103-0 (paper)

THE MEANING OF FOSSILS

EPISODES IN THE HISTORY OF PALAEONTOLOGY

Contents

Preface to the Second Edition

The preparation of a new edition of this book gives me an opportunity briefly to clarify its purpose and structure.

I have been particularly pleased that the first edition received generous comment both in journals catering primarily to palaeontologists and geologists and in those whose readers are mainly historians of science. This was a welcome indication that the book might reach the double—though overlapping—audience for which it was written. I am also grateful to the reviewers for their helpful and constructive criticisms, and it is these that prompt me to attempt some clarification of what the book *does* and does *not* attempt to do.

More than any other single point, the implications of my use of the word 'episodes' in the subtitle of the book seems to have given rise to some confusion about my intentions. Yet if I had the chance to re-write the book from scratch, I am not sure what other word I would choose, nor do I think I would adopt a fundamentally different structure for the book. Perhaps I would be more cautious, and I might substitute the word 'essays' for 'episodes', as indeed I did in the concluding passage of the book (p. 264). Yet although the five chapters are in a certain sense separate 'essays' in the history of palaeontology, the term 'episodes' does serve to emphasise that the chapters are concerned with successive *periods* in that history. In other words, one of my primary aims was to try to convey 'a sense of period' to the reader, a sense of the intellectual and social coherence of the theories and activity of studying fossils, within each of my chosen periods of history.

On the other hand, this emphasis on the coherence of each period does perhaps suggest a greater degree of discontinuity *between* the 'episodes' than I ever intended. Such an impression is perhaps heightened by the fact that I begin each chapter with a precisely dated event. The purpose of this literary device, however, was merely to anchor each 'episode' to a concrete moment in history, to act as a starting-point—in an expository rather than a strictly chronological sense of the term—for my description and analysis of each period. More seriously, my use of discrete 'episodes' may have seemed to imply a concept of discontinuity in history with some affinity to Michel Foucault's work, and it might be thought that my 'episodic' treatment therefore tends to evade the problems of accounting for the conceptual continuities and developments from one period to the next.

In fact, however, my use of 'episodes' had no such profound historiographical meaning, and was more a result of practical constraints which even now, five years after I completed the book, are scarcely less pressing. In a book intended for a broad range of non-specialist readers, I did not think it appropriate to engage in polemical debate on the many points on which historians of science currently differ in their interpretations. I still think it more helpful to present my own interpretation in a coherent and straightforward form, so that the non-specialist reader can at least see how the picture looks to *one* historian, however personal a view that may be; while the specialists will have no difficulty in seeing for themselves how my approach relates to those of other historians.

The practical constraints to which I have just referred are concerned, however, less with controversies among the steadily growing number of historians of the earth sciences, than with the limitations imposed by the amount of research which this small band of scholars has so far produced at modern professional standards. Many years ago, at an early stage in planning the lectures on which this book is based, I discovered the unreliability of the most obvious 'secondary' works when matched against a reading of the primary sources, and I decided to restrict my lectures to topics on which I could find time to make a first-hand study of primary material. Although the general quality of published historical work on the subject has improved vastly in the last ten or fifteen years, I decided when writing this book to retain this emphasis on primary sources. I realised however that this involved putting less weight on some important aspects of the subject which other historians have explored more thoroughly, than on other aspects where I could feel the reassurance of my own first-hand knowledge of the sources.

More positively, however, I would now want to clarify my intentions by describing the five chapters of this book as a series of 'essays' dealing with certain periods in the history of palaeontology from different points of view. Within the constraints of the time that I had available for the relevant research on primary sources, the essays were and still are historiographical experiments of two or three distinct kinds. The shifts in the mode of historical writing through the book, which some reviewers noticed, are also a fairly direct consequence of the historical development of palaeontology itself. I mentioned this point briefly in the preface to the first edition, but it is perhaps worthwhile to give it more emphasis now.

The first two chapters deal (broadly) with the sixteenth and seventeenth centuries, before 'geology'—let alone 'palaeontology'—had become a perceptible or definable field of knowledge. They were periods, however, when the basic phenomena for the future science of palaeontology, namely the peculiar objects called fossils, were already matters of debate among a growing circle of 'naturalists' and 'natural philosophers'. It is obviously appropriate to describe and analyse the interpretation of fossils in these periods within a very broad historical framework, rather than as a narrowly technical problem. I have therefore tried to relate 'the fossil problem' to the general intellectual currents of the time, and to indicate the social context in which fossils were studied and discussed. At the same time I have tried to point out what historians of science tend to underestimate, namely the intrinsic constraints on interpretation that were imposed by the kind of evidence—in other words the particular fossils—that happened to have been discovered at any given time. Although I begin each of these first chapters with a precisely dated event, these are—as I have already explained—simply convenient 'markers' to anchor my interpretation to concrete moments in history. In reality, the first two chapters form a connected whole, and have an underlying narrative form which brings the history of 'the fossil problem' from the mid-sixteenth century (or before) as far as the early years of the eighteenth. My interpretation could undoubtedly be enriched by more attention to the evidence of the less formal writings of the individuals I mention—for example their letters—but within the time that I had available I felt it legitimate to concentrate on what they chose to make available in published form; a tip of the iceberg perhaps, but a tip which is nevertheless continuous with what lies beneath.

After the second chapter, there is a historical lacuna that comprises most of the eighteenth century. I certainly do not (and did not) subscribe to the view that in the earth sciences the eighteenth century was altogether a period of stagnation or even decline. That view can now be seen to be partly a product of early nineteenth-century propaganda, by geologists who devalued their predecessors in order to legitimate their own self-consciously 'new' science. Recent historical research is making it steadily clearer that the eighteenth century was a period in which many of the fundamental features of later geology, such as fieldwork, collecting, and the attempt to construct theories that could be related more closely to observation, were first firmly established. But two practical constraints forced me to treat that century almost as *terra incognita*: relatively little has yet been published on the subject at modern scholarly standards, and I myself quickly discovered that even a superficial study of the relevant eighteenth-century primary sources was more than I had time for. At this point in the book, my use of the term 'episodes' must regrettably be regarded as an excuse for not giving a full account that bridges the gap between the first and last years of the eighteenth century; but I hope that others will fill that gap in due course.

By the time that I take up the story again, at the beginning of the third chapter, we have reached a period at which the study of fossils had become enormously enlarged and deepened, whether considered as an activity of certain individuals in society or as a body of observations, concepts and theories. The ever-present constraints of time and space here forced me to make a decision, which I do not regret, to focus attention on the *conceptual* development of the study of fossils. In this respect the third and fourth chapters form a connected narrative ranging from the last years of the eighteenth century to the middle of the nineteenth. I have tried to indicate the social context in which the theorising of this period was done, and its implications for other intellectual and social currents; but unavoidably these are treated mainly as a 'background' to the development of the *theories* of palaeontology.

I felt that the most useful contribution that I could make, as a palaeontologist turned historian, was to emphasise the interaction between theory-building and the accumulation of ever-richer stores of evidence —namely the fossils themselves—during this period. For historians of science who incline towards abstract historiography, like their cousins the philosophers of science, tend to take their examples from precisely those natural sciences where this intrinsically cumulative element is absent or inconspicuous; and I felt—and still feel—that it is worthwhile to bring other equally interesting sciences to their attention, and to try to convince them that a science like palaeontology is not to be dismissed as mere 'stamp-collecting'.

Like every decision between attractive alternatives, my relatively 'internalist' treatment involves giving less emphasis than I would wish, to other equally interesting aspects of the period. It would certainly be possible, legitimate and important to trace in more detail the development of the study of fossils as a social activity, making use not only of institutional sources but also the informal evidence of correspondence and field notebooks. But I felt it more important to concentrate again on the *public* debate that can be traced in published works, although I am well aware of the further insights that unpublished material can provide. Moreover, even within my chosen approach I had to restrict my discussion to the main theme of the use of fossils in the development of 'high theory', in other words in the construction of a high-level picture of the history of life and its high-level interpretation.

This necessary focus on certain aspects of the conceptual development of palaeontology applies with increased force in the fifth and last chapter. I have there argued that by the later nineteenth century the science of palaeontology—the term is here at last no longer an anachronism—had acquired most of the characteristic features of its modern form. The sheer volume of published work alone, although unremarkable by late twentieth-century standards, is enough to force still further specialisation on the historian too, if banal generalisations are to be avoided. I decided to concentrate in this final chapter on those aspects of palaeontological work which were seen at the time as bearing on the great questions of 'high theory' in biology; and in particular therefore, on the development of evolutionary theory. This is not, however, just another example of what might be called the 'All roads lead to Darwin' syndrome in the historiography of nineteenth-century science. Of course many palaeontologists of this period were uninterested in evolutionary theories and simply went on with routine (though often highly sophisticated) research into the taxonomic and stratigraphical aspects of their science. But I think that my focus on the palaeontological contribution to evolutionary theory is historically justified in the sense that most of the leading scientists concerned—both palaeontologists and those who looked at the science from the outside—saw this as its most important contribution to the general scientific understanding of the natural world.

I do not want to end this preface on a defensive note. I have tried to explain how the various chapters of this book try to do rather different things. It should go without saying that I welcome other treatments of the subject which use different approaches. Nevertheless I hope that my own treatment, in this new edition, will continue to give historians and palaeontologists a useful insight into some of the problems that the study of fossils has posed in the past, and that it will stimulate some readers into further research on the subject.

For technical reasons it has not been possible to make major changes in the text of this edition, or to add full references to recent publications, but I have taken the opportunity to correct a number of misprints and one or two small factual errors which correspondents have kindly pointed out to me. Since I did not see the original illustrations in page proof, one of them (Fig. 1.11) was printed upside-down, and this has now been corrected. I am grateful to Mr. Neale Watson and his staff for their care and attention to the preparation of this edition, and particularly for endeavouring to improve the quality of reproduction of the illustrations.

Amsterdam
April 1976

Preface

FOSSILS are often objects of striking form; some are spectacular, and many have aesthetically pleasing shapes. It is therefore not surprising that some fossils have been noticed and commented on by men of many different periods and cultures. However, it is only within Western civilisation, during the period since the Renaissance, that palaeontology has emerged from this diffuse awareness of fossils and developed into a coherent scientific discipline. This book is an essay towards the understanding of that emergence and development. However, the subject is far from being merely peripheral to the wider story of the development of modern science as a whole. A knowledge and understanding of fossils played a crucial rôle in the recognition of the immense age of the Earth and in the development of evolutionary theory. In these ways palaeontology has influenced fundamentally our conception of the natural world and of our own human place within it.

This book had its origin in a course of lectures which I gave at Cambridge several years ago at the invitation of the History and Philosophy of Science Committee. Although they were intended primarily for undergraduates reading History and Philosophy of Science, the lectures were also attended by other natural scientists who were specialising in Geology. I was therefore aware that I had to speak to two groups with significantly different backgrounds, and I could not take for granted either an acquaintance with the technicalities of palaeontology or a knowledge of more than the outlines of the history of science. In re-writing these lectures for publication I have tried to bear in mind the two corresponding groups of readers, and I have tried to avoid assuming too much background knowledge in either field. But an exact balance between the needs of two such groups is almost impossible to maintain, and I am aware that I have tended to write more for the palaeontologists, among whom I was working when the lectures were given. Historians of science may therefore find the historical treatment somewhat elementary, but I hope this may be counter-balanced by the emphasis I have put on the practical problems of interpreting fossils in each period: in the present fashion for stressing the underlying 'non-scientific' factors in such debates, the constraints imposed by the precise evidence available at each period are liable to be overlooked. In few sciences are such constraints more important than in palaeontology, where the chance discovery of even a single specimen can (and has) become the centre of controversy on fundamental issues. To help the non-palaeontological reader a glossary of scientific terms is given at the end of the book.

Although the completion of this book has been unavoidably delayed I have kept much of the essential framework that I used in the lectures. I have also retained something of their style, if only to emphasise that this book makes no pretence at being an exhaustive or 'definitive' account of the history of palaeontology. The historical value of detailed accounts of single branches of science would be questioned—and rightly in my opinion—by many historians of science today; but in any case it would be impossible for such a history of palaeontology to be written at the present time. The primary source material is vast in extent and still largely unread; much of the secondary material is outdated and unreliable. I have therefore chosen to focus attention on a few aspects or episodes which seem to me to have been historically important or at least characteristic of their period, and to build a narrative around these. I have tried to set the detailed debates about fossils within a broader framework of the ideas and concerns of each period; but I believe it is a reflection of the growth and specialisation of palaeontology that my treatment of the subject necessarily becomes more technical in the later chapters.

The study of the history of the earth sciences as a whole is still in a primitive state compared, say, to the history of physics or cosmology. Much of the best modern historical work is published as articles in specialist journals, or in books that may appear to be only marginally concerned with palaeontological or even general scientific issues. Most of the more accessible works on the history of palaeontology were written within an earlier historical tradition from which the rest of the history of science has been breaking free. Adams's *Birth and Development of the*

Geological Sciences, for example, was written from a point of view all too common, even today, among practising scientists: 'These early fables of geological science', Adams wrote, 'should be read by all who are in need of mental recreation and who possess the required leisure and a certain sense of humor'. Geikie too, though far from insulting his subject-matter by assuming its triviality, expressed a similar view of the history of science in his *Founders of Geology:* science had been a progressive struggle of enlightened intellects to free human knowledge from the shackles of obscurantist attitudes; and the figures of the past could be divided, broadly, into those who had been 'correct' and those whose opinions had been 'erroneous'. To historians of science today this kind of historiography is a dead horse that is no longer worth flogging. However, since there is little modern historical work on the subject available to palaeontologists, I have felt it worthwhile to try to point out how, taken in the context of their own time, the *dramatis personae* of the history of palaeontology can by no means be divided into heroes and villains. It is of course more fruitful, and more interesting, to refrain altogether from allotting them credit or black marks for their opinions, and to try instead to understand them as men of their own time, grappling with problems which they rarely had enough evidence to solve, and solving them, if at all, in terms of their own view of the world. Indeed I believe that this is the only justification for writing a 'tunnel history' of one branch of science over several centuries: that each period's interpretation of the meaning of fossils may be an illuminating reflection of that period's view of the natural world.

I am grateful to my colleagues Dr Robert Young and Dr Michael Hoskin for valuable comments on a draft of this book, but they bear no responsibility for its final form. As an introduction to some of the historical problems raised by the study of fossils, it will have served its purpose if it encourages some readers to study more closely the episodes I have described.

Chapter One

Fossil Objects

I

ON 28 July 1565 Conrad Gesner (1516–1565), the greatest naturalist of his century, completed his book *On fossil Objects*[1]. It is an appropriate date to choose as a starting point for this history of palaeontology. Gesner's book marked a crucial moment in the emergence of the science, for it incorporated three innovations of outstanding importance for the future; but at the same time its form and contents epitomise perfectly the scientific and social matrix within which that emergence took place.

The short title of Gesner's book is deceptive: more fully it is *A Book on fossil Objects, chiefly Stones and Gems, their Shapes and Appearances.* This shows at once that the word 'fossil' has changed its meaning radically since Gesner's day. By origin the word meant simply 'dug up', and Gesner, like all his contemporaries and his predecessors back to Aristotle, used it to describe *any* distinctive objects or materials dug up from the earth or found lying on the surface. This of course included fossils in the modern sense, but it also embraced much more. Gesner's book dealt with a number of objects that we would now recognise as the fossil remains of organisms, but they were described in the context of a wide variety of mineral ores, natural crystals, and useful rocks.

This change in the meaning of the word 'fossil' is far more than a trivial point of etymology: it is a clue to the first major problem in the history of palaeontology. This was not simply to decide whether or not fossils were organic in origin. Nor was it merely a matter of recognising their 'obvious' resemblances to living animals and plants, and of combatting 'absurd' ideas that they could be anything other than the remains of those organisms. On the contrary, their resemblances to living organisms were generally far from obvious or easy to perceive; and even when perceived, it was far from absurd to suggest that those resemblances might not be causal in character.

Early naturalists such as Gesner were faced with a very wide variety of distinctive 'objects dug up'. With respect to organic resemblances, these objects can be arranged in a broad spectrum. At one end of the spectrum lie objects that had little or no similarity to organisms. Crystals such as gem-stones and useful rocks such as marble are of this character. At the opposite end of the spectrum are objects that resemble organisms so clearly that the analogy is impossible to overlook. Many fossil shells and bones are of this character. But between these extremes lies a very wide variety of objects having some degree of resemblance to organisms, but in which that resemblance is ambiguous and difficult to interpret. In modern terms this category includes many fossils with confusing modes of preservation, and others belonging to extinct groups of organisms; but it also includes many concretions and other inorganic structures with some fortuitous resemblance to organisms.

In retrospect, we can see that the essential problem was that of determining *which* of this broad range of objects were organic and which were not. It is therefore misleading to say that some early writers believed that fossils were organic whereas others did not. It is essential to discover what kinds of 'fossil' they had in mind[2]. Somewhere along the spectrum, objects with significant resemblances to organisms had to be distinguished from those in which such resemblances were either absent or purely fortuitous. However, the criteria from making this distinction were not self-evident. When in the course of time they became clearer, objects with a causally significant resemblance to organisms came to be termed 'organized fossils' or 'extraneous fossils', to distinguish them from the rest of the broad range of 'objects dug up'. But it was not until the early nineteenth century that the word 'fossil', without qualification, finally became restricted to this end of the spectrum—though even today a relic of its former breadth of meaning is still preserved in the use of the term 'fossil fuels' for coal and oil. Meanwhile the inorganic origin of many other 'fossil objects' was also becoming clearer. This left in the middle of the spectrum a gradually shrinking group of objects of uncertain origin; and in modern palaeontology this group persists under the name of *Problematica*, as a collection of objects that are doubtfully organic or at least of uncertain affinities.

The question of the nature of fossils was not, therefore, resolved in a simple struggle between 'correct' and 'erroneous' opinions: it was a much more subtle debate about the meaning and classification of the whole spectrum of 'fossil objects'.

II

Before analysing the earlier stages of the debate about 'fossils' it is worth considering the context in which they were studied by sixteenth-century naturalists. Gesner intended his small book on 'fossils' to be no more than a preliminary essay, to be followed at a later date by a full-scale work on the subject. The larger work was never written: only a few months after completing the preliminary book he died at his home in Zurich in an outbreak of plague, leaving behind him a vast mass of unpublished materials. Because his work on 'fossils' was only a small part of a much wider programme to cover the entire range of natural history, his published *History of Animals* (1551–8)[3] gives us an indication of the character that his larger work on 'fossils' would have had. In its structure and contents we can see reflected the distinctive attitudes and methods of a Renaissance naturalist, and the same features can be detected in miniature even in the small book *On fossil Objects*.

The characteristic attitude to history of the men of the Renaissance, by which they regarded their own period as a time of re-birth and attempted recovery of the values and achievements of classical Antiquity, led naturalists such as Gesner to adopt an encyclopaedic approach to their subject. To some extent this was a deliberate imitation of the classical model set by Pliny in his *Natural History*, which was reprinted many times during the sixteenth century. But it also reflected their recognition of the value of *both* the writers of Antiquity *and* their own contemporaries. Gesner's *History of Animals*, for example, was intended to be a worthy successor of Aristotle's great work of the same name; but it set Aristotle's observations alongside those of Gesner's own contemporaries. It was designed to gather together all that had been written on animals from Aristotle's time to Gesner's own, to compare and collate these opinions, and so to provide a firm foundation for future study. It seemed essential to record in full the opinions of writers ancient and modern, even though their views often conflicted with each other, and even though the compiler himself was sometimes sceptical of their more sensational assertions. However, in an age when geographical exploration was expanding the bounds of natural history almost yearly, often with the discovery of remarkable and unexpected creatures, few reports could safely be dismissed *a priori* as absurd; and Gesner thought it prudent to include many curious monsters in his compilation, while expressing doubts about their authenticity.

With this all-embracing compilative aim, it is not surprising that extremely massive encyclopaedic works were produced by sixteenth-century naturalists. Gesner's works are a good example: he published four huge folio volumes on animals; two more remained unpublished at his death, and at that time he was also working on a botanical compilation of similar scope. There can be little doubt that his fuller work on 'fossils', had he lived to write it, would have had the same character. This is suggested also by the comparable work of his near-contemporary the Bolognese naturalist Ulysse Aldrovandi (1522–1605), who in his much longer life wrote similar encyclopaedic works on all branches of natural history. His *Museum of Metals* (1684)[4]—the word has narrowed in meaning like 'fossil', and at this period meant broadly all mineral materials—remained unpublished until nearly eighty years after Gesner's small book had appeared; but in its massive bulk and encyclopaedic contents it probably resembles what Gesner's larger work might have been.

The Renaissance background of Gesner's work is shown not only in its encyclopaedic character but also in the philological emphasis of its contents. This reflects his training as a humanist scholar. His literary education, based on the classical languages, had given him a respect for the standards of exact textual scholarship that underlay the critical new editions by such scholars as Erasmus. Transferred to his work on natural history, this led him to place great importance on determining exactly what the classical authors had written about animals and plants, and to give their opinions great weight. The achievements of the naturalists of Antiquity, and especially Aristotle, had indeed been so remarkable that the respect was fully deserved. But in order to make full use of that achievement, precise identification was imperative. In his treatment of each organism Gesner therefore gave first place to questions of nomenclature and synonymy. Likewise his small book on 'fossils' mentioned the Latin, Greek and German names for the objects he was describing, and he promised that his larger work would deal at length with their "philology" (Fig. 1.1).

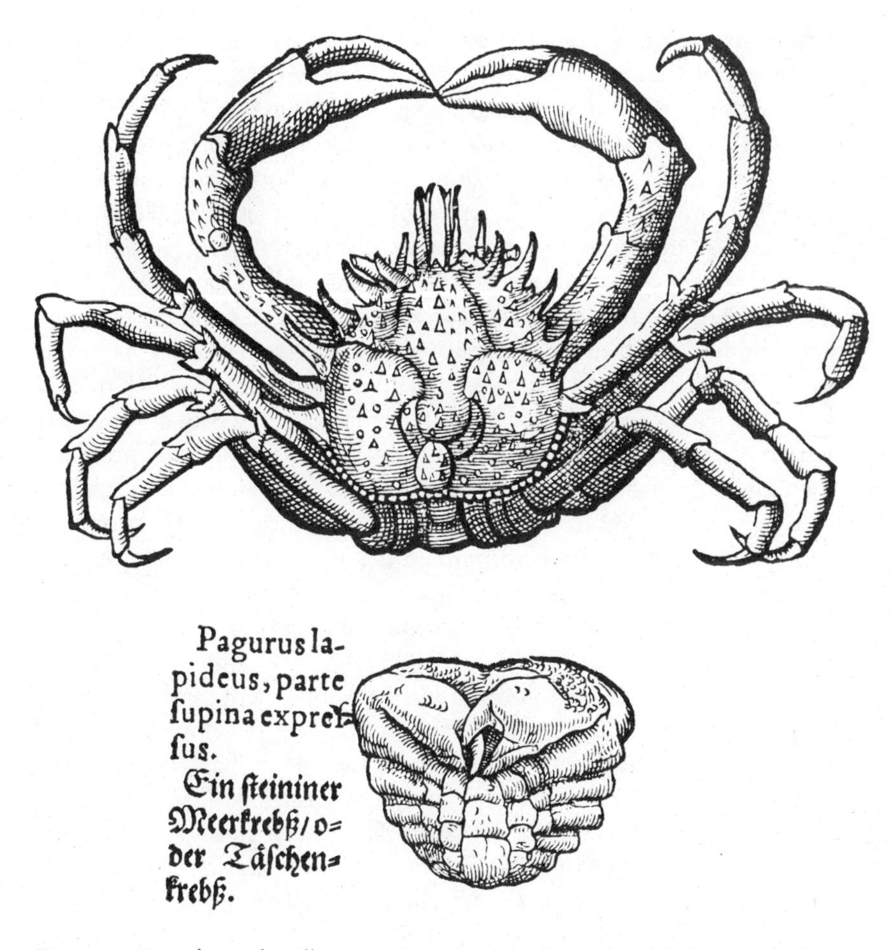

Fig. 1.1. Gesner's woodcut illustrations (1565, 1558) of a fossil crab (below), *and of the living crab ('Pagurus') which it resembled* (above). *Note the bilingual caption to the fossil*[1.15].

Gesner's concern for precise identification provides the context for the most important innovation incorporated in his book *On fossil Objects*. It was the first in which illustrations were used systematically to supplement a text on 'fossils'. The importance of this can hardly be exaggerated. Several books describing a similar range of objects had been printed earlier in the century, and some of the names used in them can be traced back through the mediaeval 'lapidaries' to the works of classical authors. However, without illustrations no writer could be certain that he was applying a name in the same sense as his predecessors. The effect of Gesner's innovation can be seen with striking effect if his book is compared with the earlier and more famous work *On the Nature of Fossils* (1546)[5] by the German naturalist Georg Bauer (1494–1555)—better known by his literary name Agricola. Both books dealt with much the same range of objects; but in the complete absence of illustrations it is often very difficult to know just what objects Agricola was describing, whereas in Gesner's book it is generally clear at once from the woodcut illustrations. Since the nature of most 'fossil objects' was poorly understood, it was difficult for any sixteenth-century naturalist to decide which features were essential for description and which merely accidental, or indeed to know how best to describe in words any features whatever. Illustrations provided a means of by-passing this problem, by allowing non-verbal communication between author and readers, and thereby mitigating the hazards of inadequate verbal means of expression. Gesner himself recognised the importance of what he was doing, for he said he was including as many illustrations as possible "so that students may more easily recognise objects that cannot be very clearly described in words"[6].

The employment of illustrations to supplement and explain a scientific text was not in itself an innovation. In more established branches of natural history the use of woodcuts had already been brought to a high standard of artistic and scientific excellence. Leonhart Fuchs's magnificent *Commentaries on the History of Plants* (1542) and Andreas Vesalius's great work *On the Construction of the Human Body* (1543), each illustrated with drawings of superb quality, had been published more than twenty years earlier[7]; and Gesner himself had used hundreds of woodcuts in his *History of Animals*, the volumes of which were a monument to the usefulness of illustrations as an aid to identification (see Fig. 1.9). For depicting 'fossil objects', however, there was virtually no precedent, no iconographical tradition to follow. One minor work published some years earlier had included a few small woodcuts, two of which can be recognised as drawings of fossil shells (Fig. 1.2); but this seems to have been Gesner's only precedent[8]. His own book was similar in size and scope, but he included a far greater number of woodcuts, providing illustrations systematically for every part of his subject matter. But even drawings of 'fossils' suffered to some extent from the same limitations as verbal descriptions, since it was not always clear which features most deserved emphasis. Gesner was aware

Fig. 1.2. Two woodcuts of fossil shells (1557), probably the first illustrations of fossils ever published in a printed book in the West[8]. They were re-published later by Gesner.

of this, saying he hoped that if his readers "find some difficult to recognise they will blame, not me, but the difficulty of the task". Nevertheless, crude though some of his woodcuts are, they initiated a technical change which was of major importance to the future science of palaeontology.

The further exploitation of illustrations as an aid in the identification of 'fossils' can be seen in the hundreds of woodcuts in Aldrovandi's book, which in this feature too is probably an indication of what Gesner's larger work would have been like (Fig. 1.3). Woodcuts, however, had their limitations: unless they were very large (as some of Aldrovandi's were) they enforced a relatively coarse style of drawing, which was ill adapted to the increasing emphasis on precise description. By the end of the century, therefore, naturalists were beginning to exploit one of the striking new inventions of Renaissance artists, namely the technique of engraving on copper. Although this was more costly, in the hands of a competent engraver it allowed far more detail to be shown, and far more subtle shading to give a greater illusion of three-dimensional solidity (Fig. 1.4). In this respect Aldrovandi's book on 'fossils' was already old-fashioned by the time it made its belated appearance; copper engravings had by then been used for more than thirty years for illustrating fossils, some of the first (see Fig. 1.11) having been published early in the seventeenth century by the Neapolitan naturalist Fabio Colonna (1567–1650)[9]. The change from woodcuts to copper engravings was only the first of many technical advances in illustration, generally taken over from the visual arts, by which palaeontologists have been able to improve the quality and precision of their non-verbal communication with each other. This dependence on

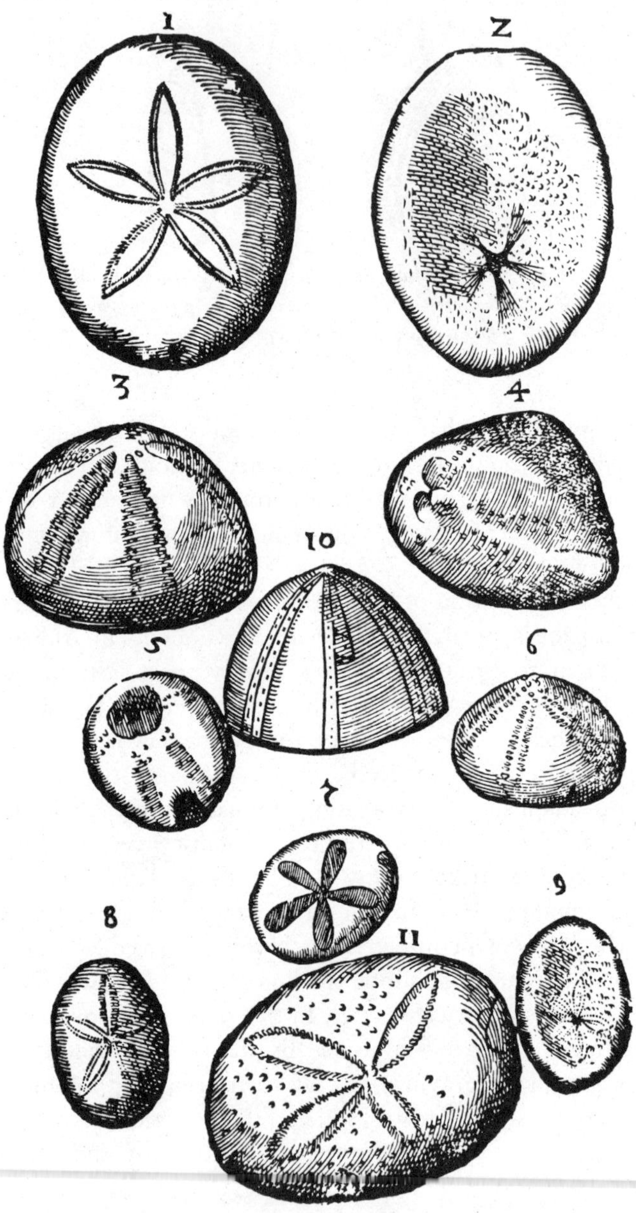

Fig. 1.3. A page of illustrations of fossil sea-urchins from Aldrovandi's large book of fossils[4]. Contrast the crudity of these 16th-century woodcuts with the delicacy of an early 17th-century copper engraving of a similar fossil (Fig. 1.4).

Fig. 1.4. A copper engraving of a fossil sea-urchin, from a museum catalogue published in 1622[12]. Compare with the relatively crude treatment of similar fossils in earlier woodcut illustrations (Fig. 1.3).

illustrations is not a reflection of the 'immature' state of the science, but is an essential element in its structure, stemming from the inherent nature of its subject-matter. Technical advances in illustration might be said to have played a part in the history of palaeontology similar to that of improvements in instrumentation in the physical sciences.

III

Gesner's use of illustrations in natural history reflects not only his concern to identify precisely the material described by the Ancients, but also his emphasis on the importance of first-hand experience. His respect for the opinions of classical authors was tempered by a method of study in which great weight was placed on the value of personal observation. In all his work on natural history, Gesner compiled his material as far as possible from a basis of first-hand observation, or, where that was not possible to him, at least from a study of preserved specimens. Indeed he followed Fuchs's example of employing a draughtsman and an engraver to make illustrations under his direct supervision, in order to ensure the highest standards of accuracy in representing the specimens he collected or was sent.

This emphasis on the importance of looking at nature for oneself is characteristic of a strand in sixteenth-century thought which was, to some extent, the opposite of the humanist emphasis on recovering accurately the writings of the Ancients. Advances in technology and voyages of exploration were beginning to provide a 'model' of human history that would turn attention away from an exclusive concern with recovering a golden past and would begin to persuade men that their own age might even surpass that of Antiquity. This feeling, especially among those most closely involved in practical pursuits affected by new discoveries, encouraged the view that nature too should be studied without uncritical regard for the opinions of the Ancients. Among those who wrote on 'fossils' the French ceramic craftsman Bernard Palissy (1510?–1590) is a good example of this anti-traditional tendency. Palissy's travels as a 'journey-man' potter brought him first-hand experience of a wide range of 'fossils', and especially of the materials for ceramics. At the same time he was proudly ignorant of the classical languages and the traditional teaching of the universities, and took a delight in exposing the supposed errors of more 'learned' writers.

This anti-authoritarian outlook has sometimes been linked directly with the Protestant rejection of Catholic tradition. For 'mainstream' Protestants, however, the importance of personal experience was always balanced by an emphasis on the centrality of the bible, which gave them a natural affinity with the wider humanist movement. Humanist scholars were concerned to recover not only the writings of classical Antiquity but also the biblical documents, which were equally a legacy from the ancient world and therefore equally amenable to the same textual methods: in both cases the aim was to get behind the accumulated corruptions of more recent centuries to the purity of the original texts. Gesner, who was born and lived most of his life in Zurich, one of the centres of Reformation thought, would have felt sympathy with this task: he had learnt not only Greek but also Hebrew in order to read the biblical documents in the original. It has indeed been argued that the Protestant concern to return to the original sources of Christianity positively encouraged a comparable concern to study at first hand the whole world of nature, and Gesner makes a good case for this thesis[10]. There were, it is true, more radical currents of thought, in natural science as in religion, that rejected the authority of tradition more emphatically; but Gesner was too well aware of the value of the Ancients to follow such a path in natural history, just as his close friend the Zurich reformer Heinrich Bullinger was too well aware of the centrality of the New Testament to follow it in theology.

The encyclopaedic approach, combined with the emphasis on personal observation of nature, provide the context for the second innovation incorporated in Gesner's book *On fossil Objects*. The basis for this descriptive work was the formation of a collection of specimens. Published illustrations were, in effect, merely a convenient substitute for a museum: they could be duplicated in large numbers by printing, thereby placing the same data at the disposal of naturalists everywhere. However, even illustrations could be misleading and ambiguous, and their value was greatly enhanced if, in cases of doubt, it were possible to study the original specimens from which they had been drawn. This involved the deliberate formation and preservation of museum collections. Here again, as with illustrations, more established branches of natural history had already led the way. Botanic gardens were founded at many universities during the sixteenth century, and were supplemented by the invention of the 'dry garden' (*hortus siccus*) for pressed plants. Animals were more difficult to preserve, though at least skeletons and shells could be collected. The formation of museums containing materials for natural history grew naturally out of the Renaissance enthusiasm for collecting the relics of Antiquity; and in many early museums all kinds of object, natural and artificial, were assembled haphazardly.

For 'fossil objects', however, museum preservation was still more suitable than for animals and plants: it was not merely a useful but inferior substitute for an assemblage of living organisms, but rather, like a collection of antiquities, the best possible means of preserving the objects concerned. Agricola and other earlier writers may well have formed collections of their own, but Gesner's book is the first work on 'fossils' that clearly refers to such a collection. Gesner expressed his gratitude to his friend the physician Johann Kentmann of Torgau (1518–1574) for sending him specimens to supplement his own, and he repaid the debt by placing the catalogue of Kentmann's collection at the front of the composite volume in which his own work was bound[11].

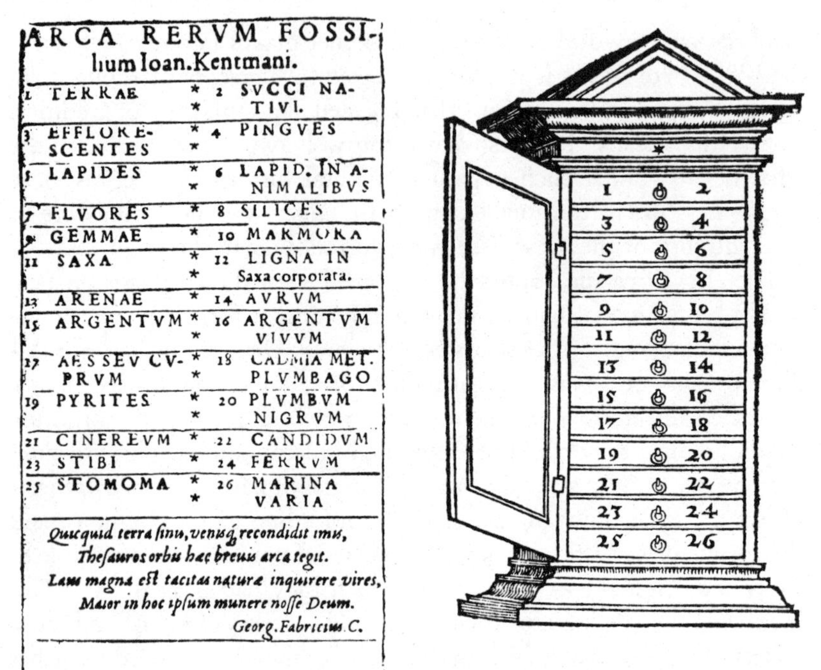

Fig. 1.5. The first published illustration (1565) of a museum collection of 'fossil objects:' Johann Kentmann's "Ark" or cabinet[11]. Note the wide range of objects indicated on the key: for example "earths" (1), flints (8) and marbles (10); gold (14), silver (15) and iron (24); 'stones in animals' (6) and 'various marine' objects (26). Most fossils in the modern sense would have been included under 'stones' (5) and 'wood embodied in rocks' (12).

The importance of the museum as an innovation in this branch of natural history is symbolised by the frontispiece of Kentmann's catalogue—the only illustration it contained. His little cabinet with its numbered drawers (Fig. 1.5) was termed significantly an 'ark' (*Arca*), which emphasises its function for the preservation of 'fossil objects'. Without the establishment of a tradition of museum preservation, it is difficult to imagine how a science of palaeontology could have emerged. As with the use of illustrations, the importance of museums is not a sign of the immaturity of the science, an indication of a 'descriptive' phase not yet outgrown: on the contrary museums are a necessarily central feature of the activity of studying fossils, stemming again from the inherent nature of the material.

Kentmann's 'ark' and his published catalogue of its contents were soon followed by similar but grander schemes. A 'Room of Minerals'

(*Metallotheca*) was established in the Vatican, for example, to parallel the papal Library (*Bibliotheca*), and its contents were described in a catalogue by the papal physician and naturalist Michele Mercati (1514–1593), and later supplemented by the botanist Andreas Caesalpino (1519–1603). The naturalist Francisco Calzolari (1521–1600) likewise formed in Verona a large but miscellaneous museum of natural history, which was continued by his son; the published catalogues again give a good impression not only of the range of its contents but also of the atmosphere of such museums[12]. Catalogues of this kind established a continuing tradition in the publication of fossils and related objects, which reaches of course to the present day.

IV

Gesner's debt to Kentmann is only one example of the pattern of activity that underlay all his work. His compilations depended not only on the previously published work of other authors, both ancient and modern, and on his own personal observations, but also on his ability to draw on new and unpublished information supplied by his network of scientific correspondents. He was not unusual in having many correspondents: scholarly letter-writing, as a revival of a classical tradition, was another way in which humanistic education directly benefitted the study of nature in the Renaissance period, besides providing in the Latin language a means of easy communication across national boundaries. In an age when travel was not to be lightly undertaken, contacts between scholars by means of letters were of course valuable for the exchange and stimulation of ideas in any branch of learning. They were, however, particularly important for a different reason in any branch of natural history. The subject-matter of astronomy or chemistry was not closely dependent on locality, whereas the study of botany or zoology necessarily was. It was therefore no accident that in the course of compiling his *History of Animals* and collecting material for a comparable botanical work Gesner had built up a network of correspondents of exceptionally wide extent, ranging geographically from Italy to England and from Poland to Spain, and crossing all the ideological and political frontiers of the divided Europe of the Reformation period.

This is the context for the third innovation incorporated in Gesner's book *On fossil Objects*. It is the first such work in which there is a clear expression of a programme of co-operative research on 'fossils'. Gesner had already received specimens and drawings from Kentmann and several other correspondents, but his book was explicitly designed to elicit further information of the same kind. It had been written, he explained, 'rather hurriedly and without much preparation', in moments of leisure from other work, specifically in order to stimulate interest in the subject. It was designed 'to encourage other students of these objects in other parts to send me more examples of stones worth recording and suitable for accurate reproduction'. An indication of the wide distribution that this request received is given by the fact that half a dozen copies of Gesner's book are preserved in the libraries of Cambridge alone (most of them having been acquired at the time), although England was rather on the periphery of the world of sixteenth-century natural science. The very publication of Gesner's book in a preliminary state is thus a reflection of the research programme he hoped it would initiate.

The importance of this innovation, like the use of illustrations and the formation of museum collections, is difficult to exaggerate. While the study of animals and plants was certainly dependent on locality, the study of fossils was and is even more so. Most animal and plant species can be found in appropriate habitats over fairly wide areas, but even the commonest fossils generally have to be collected from extremely restricted localities—a particular limestone quarry, for example, or temporary excavations for the foundations of a particular building—which may not be known or accessible to any but those living close by. More than other branches of natural history, therefore, the study of fossils requires the cooperative efforts of many naturalists living in different places.

It is symptomatic of the sense of scholarly community felt by most sixteenth-century naturalists that Gesner dedicated his book on 'fossils' not, as was usual, to some local dignitary or princely patron, but to a Polish scholar whom he knew only by correspondence and had never met. The same sense of community is reflected also in the form in which he published his book. He collected and edited seven shorter works on related subjects by other authors, and published these with his own book as a volume entitled *Several Books on all Kinds of fossil Objects*[13]. By this means he was able to publish a short work by a fellow-naturalist who had died at a tragically early age, to give wider circulation to some previously published works by other scholars, and to lend the prestige of his own name to two works by his friend Kentmann. With characteristic modesty he placed his own book last in the collection, although it was the most substantial.

It was not until the second half of the seventeenth century that the correspondence networks of Renaissance scholars such as Gesner began to be formalised into scientific societies issuing printed news-sheets to keep their members in touch with one another. But the embodiment of the scientific community in an institutional form began, however fitfully, earlier in that century. Duke Federigo Cesi's Academy of the Lynxes (*Accademia dei Lincei*) at Rome was only one of a number of short-lived scholarly communities with high and often utopian ideals for the transformation of society. It was intended that 'colonies' of Lynxes should be established in every city, on the model of an active monastic order, to create a widespread community of scholars devoted to new ideals in natural science. In fact only one such colony was ever founded, in Naples, but to this branch of the Academy belonged Fabio Colonna, and it was in the context of this idealistic community that he carried out his work on natural history[14]. Like his famous contemporary Galileo, Colonna was proud to place his title 'Lynx' prominently on the title-pages of his publications. It is perhaps no accident that his were some of the most significant studies of fossils to be published in the early seventeenth century; for these studies foreshadow, more closely than any others, the greatly extended discussion of fossils that took place later in the century within a similar institutional setting.

V

Three important innovations are thus embodied in Gesner's small book *On fossil Objects*. The use of illustrations to supplement verbal description, the establishment of collections of specimens, and the formation of a scholarly community cooperating by correspondence—these were all innovations that had already been put to good use in other branches of natural history. However, it is only with the last work of Gesner's prolifically productive life that we can see the beginnings of their deployment in the study of 'fossils'; and the work published in the century following Gesner's death demonstrated their potential value for enlarging the scope of the discussion on the nature of 'fossils'. Before considering that discussion, however, it is important to note the motives that led Gesner and his contemporaries to study natural history in general, and 'fossils' in particular.

Firstly, the world of nature was felt to be worth describing simply because it was the product of God's creative activity. The use of the natural world to provide rationally persuasive demonstration of the divine attributes had long been established in traditional scholastic theology, and continued to be influential in the Catholic theology of the Counter-Reformation. Protestant theology, on the other hand, stressed the impossibility of reaching the true knowledge of God by the exercise of fallen reason. However, this did not lead it to devalue the study of nature; on the contrary it emphasised that the believer, while knowing God only by grace and through faith, had a positive duty to acknowledge the divine artistry of the natural world in which he was placed, and indeed to delight in it. When Gesner dedicated his volume *On the Nature of Fishes and Aquatic Animals* (1558)[15], he drew the Emperor's attention to its contents primarily as a demonstration of the marvellous works of God in the depths of the sea. Likewise in introducing his book *On fossil Objects* he became so delighted with the thought that the gems among his 'fossils' were earthly reminders of the jewelled construction of the heavenly City of God, that he had to recall himself with some reluctance to the mundane description of his book.[16] Such sentiments were no mere pious formalities: they expressed an essential part of the dynamic behind the descriptive work not only of Gesner but of many other sixteenth-century naturalists. Their interest and delight in all the varied products of God's creativity were important factors in their enlargement of the scope of natural history beyond the limits of those creatures or objects that happened to be useful to mankind. This lessening of an emphasis on man in natural history was as important, in its way, as the astronomical developments that undermined an anthropocentric viewpoint more radically in the realm of cosmology.

At the same time, however, sixteenth-century naturalists also had powerful utilitarian motives. These too had strong religious foundations, since both Catholic and Protestant theologies stressed the divinely authorised ability of man to utilise the products of the world in which he lived. A straight-forward expression of this utilitarian motive for the study of nature can be seen, for example, in Agricola's plain factual accounts of useful minerals and of mining techniques[17]. In some writers, particularly the chemists who followed Paracelsus, a practical motive was closely linked with a strongly anti-traditional and even anti-intellectual outlook[18]. Palissy, while dismissing Paracelsus with as much disdain as he ignored Aristotle, is a good example of this tendency among writers on 'fossils'. In his most important work, the *Admirable Discourses* (1580), he provocatively used a dialogue form to contrast the first-hand experience of 'Practice' with the blinkered book-learning of 'Theory'; and the full title and contents of the book stressed the practical value of the 'secrets of nature' that would be discovered by following the precepts of 'Practice'[19]. The title of his earlier work, *A true Recipe by which all Frenchmen can learn how to multiply and augment their Riches* (1563), reflects with almost embarrassing clarity the utilitarian foundation of his science[20]. The context of Palissy's references to 'fossils' in both works is in fact a much wider collection of practical information on farming methods, on water conservation and springs and wells, on an ingenious design for an impregnable fortress, and so on; while even the 'fossils' are described primarily for their practical value as materials for ceramics and other useful crafts.

Gesner's work, though far from anti-traditional, does not for that reason lack utilitarian elements. On a simple level his interest in the practical value of natural history is shown by his systematic comments on the usefulness—agricultural, culinary, and so on—of each of the animals he described; and in his book on 'fossils' the same interest is shown by his inclusion of many useful rocks and minerals. A more specific utilitarian motive however, arose from the medical context of his work. Like most other writers who have so far been labelled 'naturalists', Gesner was by training not only a humanist scholar but also a qualified physician. In the later part of his life he was chief medical officer (*Stadtarzt*) of Zurich, just as Kentmann was in Torgau and as Agricola had been in Joachimsthal and Chemnitz. The chief motive behind the botanical work of the sixteenth century was quite explicitly medical: 'herbals' such as Fuchs's were written primarily to aid the correct identification of plants with curative properties[21]. This medical purpose underlay much of the natural history of Gesner and his contemporaries.

It was strengthened and extended, however, by the increasing popularity of Paracelsan medicine. By rejecting the Galenic concept of disease as an imbalance between contrasting 'humours' of the body, and by substituting the concept of a specific bodily failure requiring an equally specific remedy, Paracelsus and his followers had focussed attention on the medicinal value of specific substances. Moreover, by regarding the physiological processes of the body as a series of chemical operations directly analogous to those taking place in the outside world, they extended the range of potentially valuable substances to include minerals and metals, as well as the long established botanical remedies favoured by more conservative physicians. Gesner himself was certainly much concerned in his medical work with the use of the newly developed technique of distillation for the extraction of active principles or 'quintessences' from a wide range of natural materials[22]; and among his contemporaries discussions of 'fossils' were often set in a similar context, as for example in the *Treatise on medical Waters and also Fossils* (1564) by the Paduan anatomist and physician Gabrieli Fallopio (1532–1563)[23]. There is little in Gesner's own book on 'fossils' about the medicinal value of the objects he was describing, but that is simply a result of its preliminary character: he promised that in his larger work he would 'fully describe every kind of stone and mineral, its power and nature and also its philology'. Once again, Aldrovandi's larger book is a useful indication of what Gesner might have produced: its section on 'stones' included descriptions of the usefulness of various stones and 'fossil objects' not only for building and other practical purposes but particularly for medicine.[24].

VI

It is significant that in the phrase just quoted, Gesner placed the 'power' (*vis*) of his 'fossils' in first place, even before their 'nature': such, the phrase suggests, was the importance he attached to their 'power'. But it would be unhistorical to assume that by this word he meant merely what we would call the medicinal value of naturally occurring chemical substances; it had a much deeper significance for sixteenth-century naturalists. Whether or not they accepted all Paracelsus's teaching, many of them were profoundly influenced by the Neoplatonic philosophy that underlay it. At the roots of Paracelsan medicine was a renewed belief in the ancient concept of an ontological analogy between Man and his external world[25]. Man was the 'microcosm', the epitome of the universe, the reflection in miniature of the structure, variety and purpose of the 'macrocosm' outside him. Every feature of the universe around him could therefore be expected to have some analogy, some token or symbol, within his being. It followed that to look for specific remedies for specific ills was no mere empirical hunch but rather an attempt to trace the implications of the fundamental pattern of nature. Indeed the whole universe of Renaissance Neoplatonism was a network of hidden affinities and 'correspondences', which might be made manifest by resemblances not only between microcosm and macrocosm but also between the heavens and the Earth, between animals and plants, and between living and non-living entities. However, this network of hidden affinities was also a network of forces and powers, of 'sympathy' and 'antipathy', which were able to act at a distance. On this basis many otherwise inexplicable phenomena, such as the attractional powers of a piece of lodestone or amber, could be given a rationally satisfying interpretation.

Within this universe the most powerful forces were those emanating from the heavenly bodies, for of all created entities the heavens occupied the most exalted position in the hierarchical structure of the Neoplatonic cosmos. Astrology in its ancient forms had long been suspect in Christian thought on account of its deterministic implications; but the 'natural magic' of the Renaissance made it acceptable once more in a subtly different form[26]. Among the documents recovered by humanist scholars had been those they ascribed to an ancient Egyptian priest named Hermes Trismegistus. Although these writings were later shown to date only from the early centuries of the Christian era, during the Renaissance period they were believed to be the work of a contemporary of Moses, an early Gentile prophet of Christ and the ultimate source of Plato's wisdom. With such an impeccably respectable origin, the 'Hermetic' accounts of the deliberate use of stellar powers or influences to produce terrestrial effects became acceptable as a basis for a new form of astrology. Instead of stressing the deterministic power of celestial forces over the fate of man, depriving him of free-will and initiative, this 'natural magic' demonstrated the ability of man to manipulate these forces to his own design and for his own ends. It was 'magic' in that it sought to operate through tapping an occult network of magical forces ramifying through the cosmos; but it was also 'natural' in that it rejected the use of demonic forces (and hence condemned the practices of witchcraft) and sought only to exploit the potentialities of a divinely created natural world. It can be argued, indeed, that it played a crucial role in the emergence of modern science, in that it provided the essential sanction for a study of nature that was closely linked with purposes of practical manipulation, and thus that it lay at the roots of the distinctively modern synthesis of science and technology[27].

Whether or not such a claim can be justified, it is certain that the study of 'fossils' in the sixteenth century cannot be fully understood except against the background of this complex amalgam of Hermetic Neoplatonism. In order to capture and exploit the powerful influences of the heavenly bodies it was necessary to identify the corresponding terrestrial entities in which those powers were concentrated in accessible form. Prominent among such entities were the precious stones, which by their colour, lucidity, brilliance and rarity seemed to reflect the ethereal qualities of the heavens. Within the context of natural magic it was therefore logical to attribute to these stones the most remarkable powers. Such powers were improved if the gems were cut or polished, for this increased their celestial quality of brilliance; and they could be enhanced still further by engraving them under the right astrological conditions with appropriate images or symbols, for this increased their efficacy in drawing down the powers of the celestial bodies with which their occult affinities lay.

The importance of natural magic as a reason for interest in 'stones can be seen, for example, in Camillo Leonardi's popular *Mirror of Stones* (1502), which was reprinted many times in the sixteenth century. A direct descendant of the mediaeval lapidaries, this brief compilation placed great emphasis on the occult 'virtues' of gemstones, and Leonardi devoted about a third of the book to the discussion of 'talismans' carved with magical images[28]. The same interest can be seen in a wider context in the equally popular encyclopaedic work *On Subtlety* (1550) by the natural philosopher Girolamo Cardano (1501–1576)[29]. This was a compendium of information on all aspects of the universe, set in a Neoplatonic framework. Its contents descended from a discussion of first principles and of the fundamental elements, through a description of celestial phenomena, to a consideration of terrestrial materials; and then re-ascended the hierarchy through plants and animals, Man and his arts and sciences, supra-human 'intelligences', and so finally to God Himself. In the section *On Stones*, Cardano accepted completely the 'virtues' attributed to gemstones, but sought to give them a natural explanation in terms of their 'correspondence' with other entities[30].

There can be little doubt that Gesner too, had he lived to write his full-scale work on 'fossils', would have described their 'power' similarly

In terms of natural magic. He had met Cardano in 1552, and was certainly familiar with his work. Moreover, among his *Several Books on all Kinds of fossil Objects* he chose to re-publish a French work on gemstones[31], in which he actually restored some astrological passages that had been censored in the original edition. This work dealt explicitly with the occult powers of gems, and its citations of Pythagoras, Plotinus and 'Hermes' reveal clearly its affinities with Hermetic Neoplatonism. It is true that Gesner was sceptical about some of the alleged powers of engraved gems, but he did not disbelieve in the reality of celestial influences on the gems themselves. Another of his *Several Books* was the Greek text and a Latin translation of a mediaeval work on the twelve stones in the breastplate of the Jewish High Priests, which had been engraved with the names of the twelve tribes of Israel; and Gesner added to this a long appendix of his own, on their identification and synonymy[32]. They were often identified with the twelve mystical stones of the heavenly Jerusalem of the Apocalypse; and since an ancient tradition had equated them with the signs of the zodiac, their mystical and cabalistic significance also had an astrological dimension. In his own contribution to the volume, Gesner actually illustrated these same twelve stones, not as arranged in the breastplate but as polished or cut gems; and he grouped this extremely powerful assemblage of stones, like a necklace, around two rings—the commonest way of wearing a stone with desirable properties. One ring was set with that most brilliant stone, a cut diamond, and the other with a stone engraved with a scarab beetle, an 'image' closely associated with the hieroglyphs that were believed to record the pristine Hermetic source of natural magic. Of all the dozens of woodcuts illustrating his book *On fossil Objects*, Gesner chose this particular one, with all its powerful associations of natural magic, to decorate the title page of the whole volume, as if to epitomise its entire contents (Fig. 1.6).

For Gesner, then, as for many others who wrote on 'fossils', the potential use of these objects in natural magic was an essential part of his motive for studying them. However it was not sharply separated either from the materially useful value of 'fossils'—and especially their medicinal value—or from their value for contemplation as the works of God. Within a divinely created cosmos in which all parts were linked through a network of hidden affinities, all these purposes were fused into a coherent whole. This explains, for example, the otherwise peculiar mixture of topics that Aldrovandi appended to his description

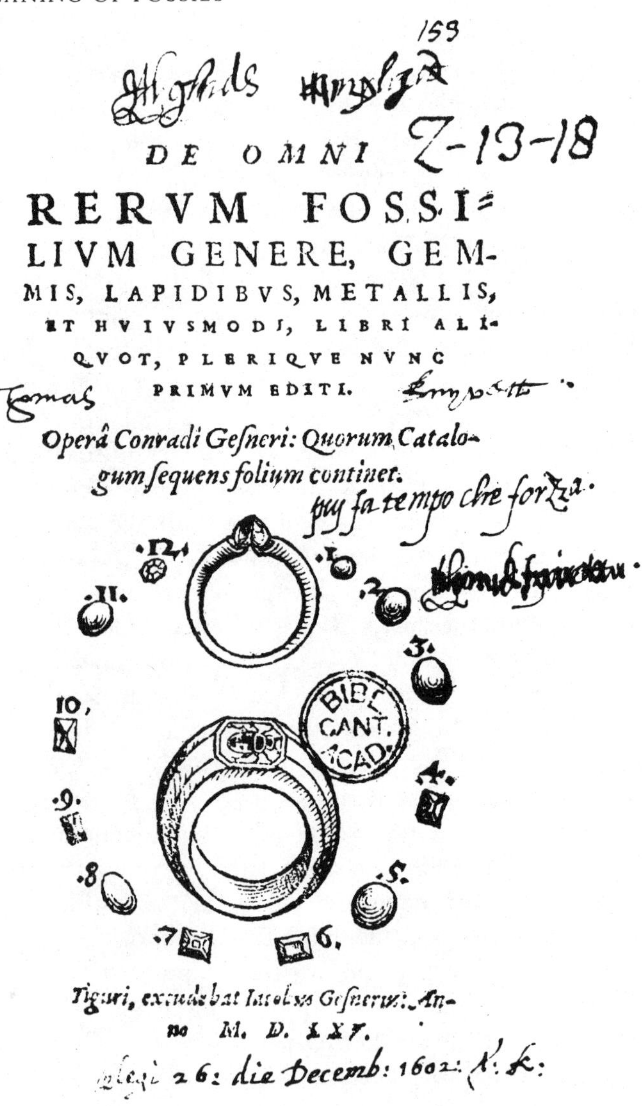

DE OMNI

RERVM FOSSI-
LIVM GENERE, GEM-
MIS, LAPIDIBVS, METALLIS,
ET HVIVSMODI, LIBRI ALI-
QVOT, PLERIQVE NVNC
PRIMVM EDITI.

*Operâ Conradi Gesneri: Quorum Catalo-
gum sequens folium continet.*

*Tiguri, excudebat Iacobus Gesnerus. An-
no M. D. LXV.*

Fig. 1.6. The title-page of Gesner's volume of works (by himself and others) on 'fossil objects' (1565)[13]. The handwriting of several early owners can be seen on this copy. For the significance of the rings and gemstones, see text.

of 'fossils': not only their practical value for building and in medicine, but also their appearances in proverbs, fables and myths, in dreams and miracles; their mystical and 'moral' significance, and their uses in pagan ritual and in prayers[33]. However 'unscientific' such topics may seem, the fact is that this is the context in which most of the descriptive work on 'fossils' was carried out during the sixteenth century; and without that work the problems of determining the causal origin of these objects could hardly have been tackled.

VII

The problems of interpreting the nature of 'fossils' did not present themselves to Gesner or his contemporaries as a matter of discriminating between the organic and the inorganic within a broad spectrum of 'objects dug up'. That is a reconstruction, made with the benefit of hindsight, of the debate as it later developed. But sixteenth-century naturalists were concerned nevertheless with the problems of classifying their 'fossils'. Some kind of grouping was an essential preliminary to any further understanding of their nature, yet no scheme of classification could avoid overtones of interpretation, if the criteria used were to be more than arbitrary.

In his work *On the Nature of Fossils*, published some twenty years before Gesner's book, Agricola had deliberately rejected the arbitrary alphabetical listing of 'fossils', which had been usual in earlier compilations. Instead he attempted to classify them primarily by their physical properties. He distinguished gems; 'earths' such as potters' clay; 'rocks' such as marble; 'metals' such as mineral ores; 'hardened fluids' such as salt, pitch and amber; and finally 'stones'. This last category included such materials as lodestone, gypsum and mica, as well as a number of objects that would now be recognised as true fossils in the modern sense. Agricola's scheme, however, also included objects that seem at first sight extraneous even to the broad meaning of the word 'fossil'. Among 'gems' he included pearls and gallstones; among 'hardened fluids', precious coral; and among 'stones', several objects that were commonly believed to fall from the heavens—though Agricola himself seems to have regarded this as no more than a superstition of the ignorant. Gesner, who always cited Agricola's work with respect, included and illustrated similar objects in his own compilation, and indeed published a separate short monograph on human gallstones, by Kentmann, in his composite volume[34]. In regarding such objects as 'fossils' Agricola and Gesner were following a long tradition; and to some extent it can be explained in terms of their interest in medicine and perhaps in natural magic. More significantly, however, what grouped all such objects with other 'fossils' was primarily their common property of 'stoniness'.

The stoniness of 'fossils' was the causal problem that was most often discussed. Aristotle had outlined an explanation in terms of vaporous exhalations, and this had been elaborated by the Arabic writer Avicenna and later by Albert of Saxony into a theory of a petrifying fluid (*succus lapidificatus*, etc.). Some such explanation was accepted by most sixteenth-century naturalists who wrote on 'fossils'. There was much to suggest that a petrifying agency was constantly at work producing stony objects of all kinds. Stalactites, for example, could almost be seen growing from the perpetually dripping water inside caves, and some springs had the uncanny power of coating objects with a layer of stone. There was a persistent belief among miners, fostered no doubt by the visible growth of secondary minerals on the walls of mine-shafts and adits, that the ores they were mining were being steadily replenished. The crystals lining the sides of mineral veins suggested that such materials as rock-crystal were being formed in the depths of the Earth just as surely as other crystals could be made to form from solutions in the chemist's laboratory[35]. Neither did the petrifying agency seem to be confined to the Earth's interior. Corals and calcareous algae showed its action in the sea, within the tissues of plants (the animal nature of corals was not discovered until the eighteenth century); gallstones and pearls proved likewise that stones could be formed within the bodies of animals and even Man himself: and meteorites and similar objects suggested, perhaps on the analogy of hailstones, that stony materials could also be produced above the Earth's surface. Objects in this last category were assumed to have originated within the atmosphere (a belief that is preserved incongruously today in the common stem of the words 'meteorite' and 'meteorology'), because within the framework of Aristotelian cosmology it was inconceivable that such irregular phenomena, like thunder and lightning, could be anything other than '*meteora*'. The petrifying agency thus seemed to pervade the entire sublunary sphere; and it seemed legitimate to study all its products together, whether they originated in, on or above the Earth.

The exact nature of the petrifying process was far from clear. Some Aristotelians described it in terms of 'vapours', others in terms of a fluid or 'juice' (*succus*). Falloppio thought there might be more than one kind of fluid, to account for the physical properties of the major classes of 'fossils', while Palissy interpreted everything in terms of percolating 'salts'. But whatever the exact explanation, such theories seemed to account, at least in principle, for the stoniness of 'fossils'.

Similarly within the Neoplatonic framework all the evidence of a petrifying process could be interpreted as signs of the 'growth' of stones. In Neoplatonic thought the distinction between living and non-living was simply unreal: all entities shared in some sense in the quality termed 'life', however much they differed in the mode of its expression. All stones, Cardano asserted, are in a sense alive, although the life of plants and animals is more manifest. But stones too clearly shared the characteristic of growth, as stalactities and crystals demonstrated. The decay of some minerals likewise suggested an analogy with disease, old age and death; and one much-discussed stone, *aetites* (probably a concretionary nodule), often contained a smaller stone within a central cavity, suggesting that it was in the act of reproduction. Even the Earth itself seemed analogous to a living organism, with percolating ground-water corresponding to the blood in the body. Whether the formation of 'fossils' was thought of in terms of some kind of precipitation from a fluid or fluids, or whether in terms of an organismic analogy of growth, the substance or *matter* of these objects seemed to be rationally intelligible. A stony matter was the natural character for 'fossils' of any kind. The *form* of 'fossils', on the other hand, seemed to be a separate problem.

The use of these categories of form and matter is partly a convenience for our analysis of the problem of fossils, for it reflects the nature of the material. In modern terms it is legitimate to distinguish questions of fossilisation, which affect chiefly the materials of which fossils are composed, from questions of biological affinity, which affect chiefly their morphology. However, this distinction, so inherently appropriate, was peculiarly congenial to the Aristotelian thought of most sixteenth-century naturalists. If the nature of any entity whatever could be analysed in terms of form and matter, the same categories could also be used to understand the nature of 'fossils'. The form of 'fossils' could therefore be studied as a problem separate from their matter.

Agricola had pointed out that many 'fossil objects' have characteristic shapes, some of which seem to imitate other objects; and he used these shapes extensively to describe his objects within the category of 'stones'. Thus for example one well-known stone, *belemnites*, imitated an arrowhead, while another, *ammonis cornu*, looked like a ram's horn. Cardano had emphasised the distinction between such resemblances, which invariably characterised certain kinds of stone, and those that were fortuitous—such as the vague likenesses that can occasionally be imagined on slabs of variegated marble. For Gesner, the Aristotelian concept of specific *differentiae* had been fundamental to all his biological work, his descriptions having been based on the reality of discrete 'species' of animals. In his preliminary survey of 'fossils' he therefore applied this concept of 'species', developing the hints that Agricola and Cardano had given, and made the *form* of 'fossils' the basis of his classification. The shapes of 'fossils', he said, like those of plants, 'are (so to speak) specific, and always appear to be attached to a certain class of object as though peculiar to it'. Moreover, the 'images' shown by these objects were all the finer, he said, in that they were clearly *natural* and not, like the Hermetic hieroglyphs, carved by merely human agency.

Gesner, as eclectic in his natural philosophy as in his sources for natural history, then grouped these Aristotelian units into classes that are reminiscent of Cardano's Neoplatonic survey of the cosmos. The 'fossils' were classed according to their resemblances to objects in other realms of nature; and the classes were arranged according to the position of their analogues in the hierarchical cosmic scheme, descending through that hierarchy (as Gesner pointed out) just as the soul in the opposite direction aspires to ascend towards God. Thus he began with stones having shapes related to geometrical figures or to the Aristotelian elements, the most fundamental entities in the universe, and then descended through those resembling the heavenly bodies and those related in some way to the realm of *meteora*, to those resembling terrestrial objects. These in turn included stones with a resemblance to the works of Man himself (and, by way of parenthesis, those objects actually owing their form to human workmanship, such as cut gemstones and engraved medallions), and so finally to those resembling various kinds of plants and animals.

VIII

With the final classes in Gesner's scheme we have at last arrived at objects which, as a glance at his woodcuts shows, fall within the modern definition of the word 'fossil'. However before considering the interpretations that sixteenth-century naturalists placed on such organic resemblances, it is worth noting that a number of fossils in the modern sense were *not* included by Gesner in his classes of objects resembling organisms. This reflects the inherent difficulties that faced any naturalist, even to the end of the following century, in perceiving organic resemblances in fossils.

In the first place, there were difficulties arising out of the modes of preservation of fossils. The existence and severity of these problems may not be apparent even to the modern palaeontologist, because he may have successfully forgotten his initial difficulties in recognising and understanding the vagaries of fossilisation. For example, it is relatively easy to recognise the organic nature of many fossil shells of geologically recent origin. They are likely to come from unconsolidated sediments, from which they can be extracted complete and in good preservation; and they are probably almost unchanged in substance, apart from the loss of their original colouring. Most fossils, however, are much more difficult to interpret. The confusing diversity of their modes of preservation can make their nature anything but obvious. Even quite ordinary fossil molluscs can be very puzzling, if for example the actual shell has been dissolved away, leaving only an empty hollow with a 'negative' cast and mould in some compact rock; or the shell itself may have been recrystallised into a sparry material quite different in appearance from the original, and with a confusing similarity to crystalline materials of purely inorganic origin.

Extinct molluscs such as ammonites (Agricola's *ammonis cornu*) can be still more puzzling, since they are generally preserved as casts in a hard rock or in crystalline calcite, or with the shell replaced by another material such as the metal-like pyrite, or as paper-thin impressions flattened on the surface of a shale (Fig. 1.7). For early naturalists such difficulties were aggravated by the fragmentary preservation of many of the commonest fossils. The organic nature of belemnites, for example, would have been easier to perceive if they had been known in more

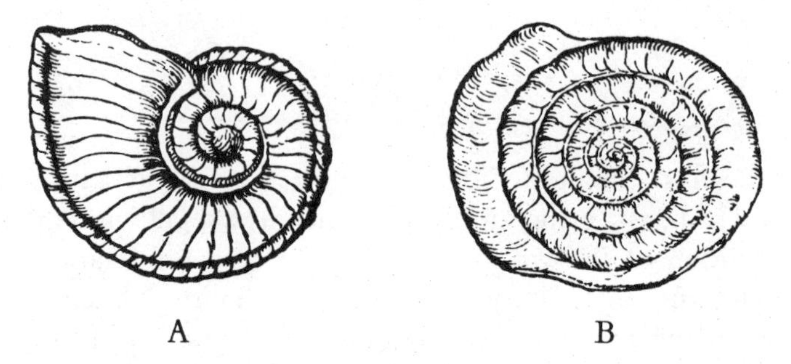

Fig. 1.7. Two woodcut illustrations of ammonites from Gesner's book (1565)[1]*. One of them (A) he placed in the same class as various fossil shells, but the other (B) he thought was snake-like.*

complete preservation: as it was, ordinary specimens preserving no more than the solid crystalline 'guard' were bound to seem closely similar to stalactites and other inorganic objects of comparable structure (Fig. 1.8, A). Likewise, complete specimens of fossil crinoids ('sea-lilies') are obviously organic in appearance, even if they look deceptively plant-like; but until such relatively rare specimens were discovered, the far more common fossils representing isolated pieces of their stems or single stem-ossicles were naturally very difficult to interpret (Fig. 1.8, B, C). Indeed this difficulty was greatly increased by their mode of preservation, for their normal calcite cleavage gave them a crystalline and therefore suspiciously inorganic appearance.

Conversely, where there *was* an obvious similarity in form between a fossil and some living organism it might be—in modern terms—purely fortuitous. One flint nodule, for example, might resemble a human foot just as closely as another resembled a sea-urchin: both objects would be reminiscent of an organic structure, without having an exact identity. We today may be sure that the resemblance is fortuitous in the first case and causally significant in the second, but without a clear understanding of the processes of fossilisation that conclusion would be far from obvious.

A second source of difficulty in perceiving organic resemblances in fossils arose out of the unfamiliarity of the organisms from which many fossils had originated. Here again the difficulty was minimal for some geologically recent fossils, for these were often closely similar to, if not identical with, living species. For many of the commonest fossils,

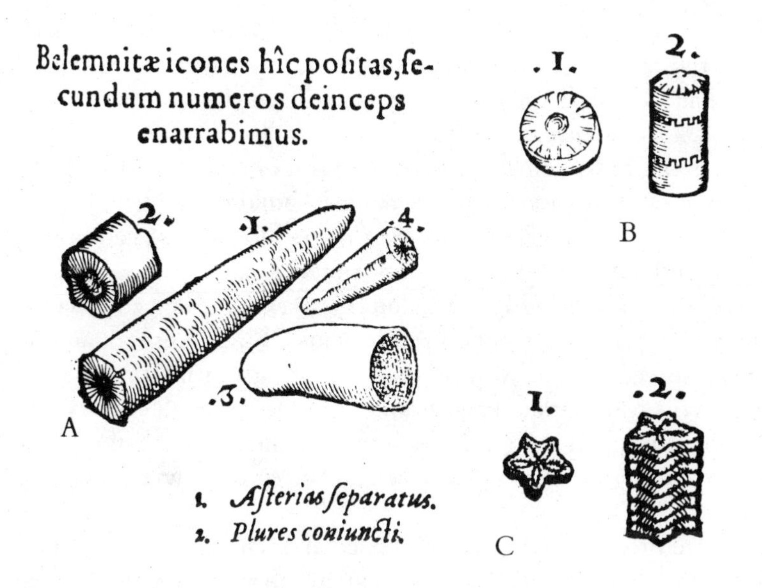

Fig. 1.8. Gesner's woodcut illustrations (1565)[1] *of belemnites (A) and crinoid ossicles (B, C): he knew no comparable living animals.*

however, the difficulties of interpretation were aggravated by the fact that they belonged to almost or totally extinct groups. Belemnites, ammonites and crinoids are all good examples of this. Belemnites, being completely extinct, had no close living analogues whatever (their closest common relatives, the cuttle-fish, have no obvious similarity); living cephalopods with chambered shells like those of ammonites were not known until the seventeenth century; and living stalked crinoids were not discovered until the middle of the eighteenth. The perception of organic resemblance in a given fossil was therefore dependent, in part, on the contemporary state of biological knowledge of its living analogues.

The effect of these difficulties can be illustrated particularly clearly from Gesner's work, because we can gauge accurately from his *History of Animals* just how many living animals he was familiar with, and we can compare these with the fossil specimens illustrated in his work *On fossil Objects*. Like many of his predecessors he was clearly aware of the organic similarities of fossil wood and fossil bones and teeth, and he placed these objects in chapters devoted, respectively, to objects resembling trees or parts of trees, and to objects resembling parts of quadrupeds. Most of Gesner's objects that are fossils in the modern sense are the remains of marine animals, and can be compared with what he had described only seven years earlier in his volume *On the nature of Fishes and aquatic Animals.*

This was the most comprehensive work on aquatic animals ever to have been compiled. Gesner had been able to build on the firm foundations of Aristotle's excellent work on marine biology; he had drawn on the recently published works of the French naturalists Guillaume Rondelet (1507–1566) and Pierre Belon (1517–1564)[36]; he had received drawings and specimens from correspondents all over Europe; and he had spent some time in Venice studying at first hand the animals brought to the fish-market. Few naturalists in the sixteenth century had a wider knowledge of marine biology than Gesner, or were in a better position to recognise the organic resemblances of a wide range of common fossils.

It is therefore not surprising that Gesner included in his chapter *On Stones resembling aquatic Animals* a fair number of objects that we would regard as correctly placed in that category. For example he had one of the fossil fish from the Permian Kupferschiefe of Eisleben in Saxony: it was a complete specimen and was obviously fish-like, although it was flattened on the shale surface and preserved curiously 'with coppery scales'. He also recognised that the objects traditionally termed *glossopetrae* or 'tongue-stones' resembled the teeth of sharks and dogfish, although they were much larger: indeed he had illustrated and commented on this resemblance when describing sharks in his biological volume (Fig. 1.9). Various mollusc shells, of both gastropods and bivalves, likewise gave him no trouble, for he was familiar with a wide variety of living molluscs; and he saw the clear resemblance between one of his fossil specimens and a crab that Rondelet had described (Fig. 1.1).

Many of his specimens, however, were more difficult to interpret. Rather surprisingly, he successfully recognised that a flint cast of an echinoid resembled one of his living sea urchins stripped of its spines and shell (Fig. 1.10 A, C): a remarkable triumph over the difficulties of fossilisation. On the other hand all the sea-urchins he was familiar with were kinds with small fine spines (Fig. 1.10 B), and he therefore failed to recognise the fossil remains of cidaroids with very large club-shaped spines. These spines (Fig. 1.10 F), which are usually

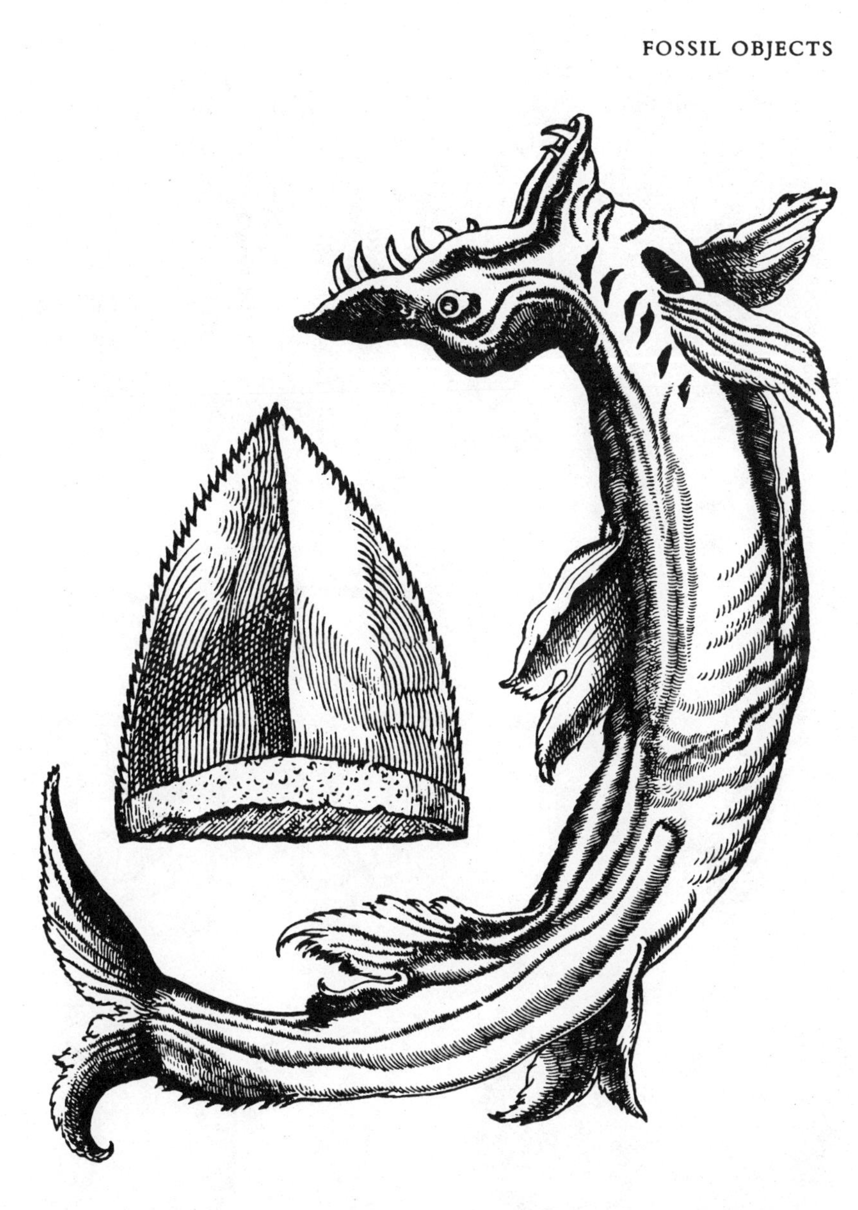

Fig. 1.9. Conrad Gesner's illustrations (1558)[15] *of a* Glossopetra (*inset*), *and of the shark whose teeth it resembled. This is probably the earliest Western illustration of a fossil described in relation to a living animal.*

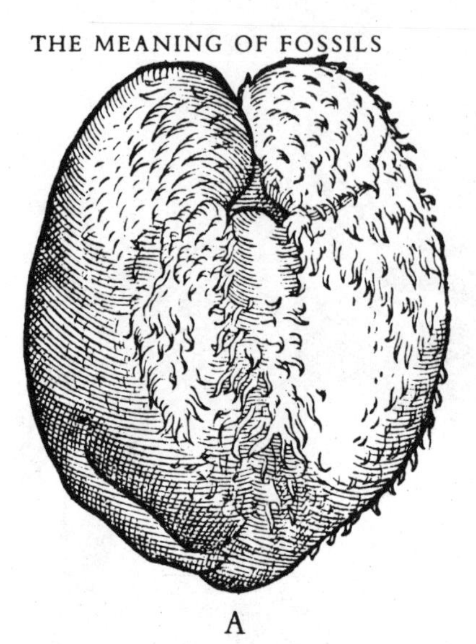

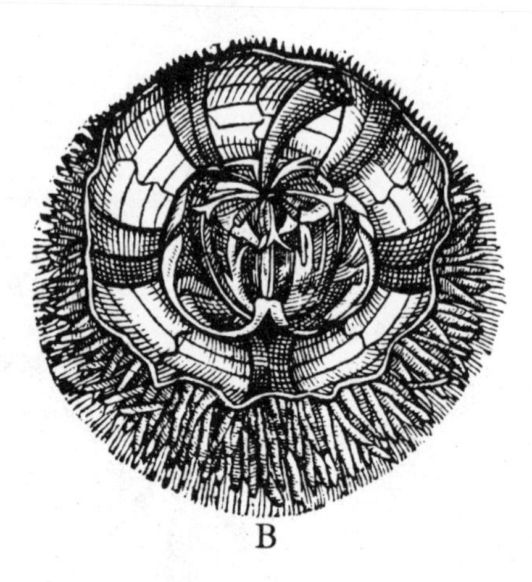

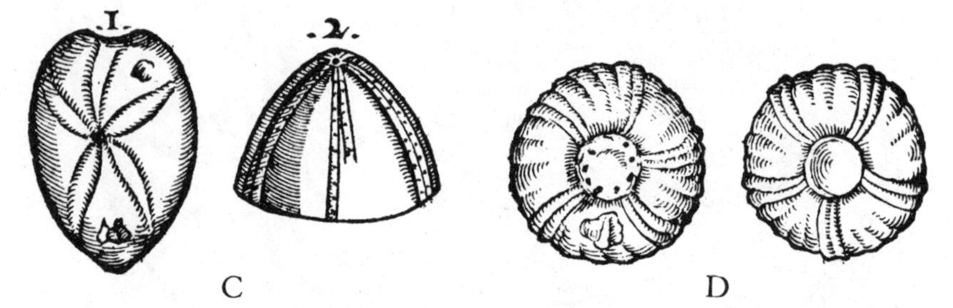

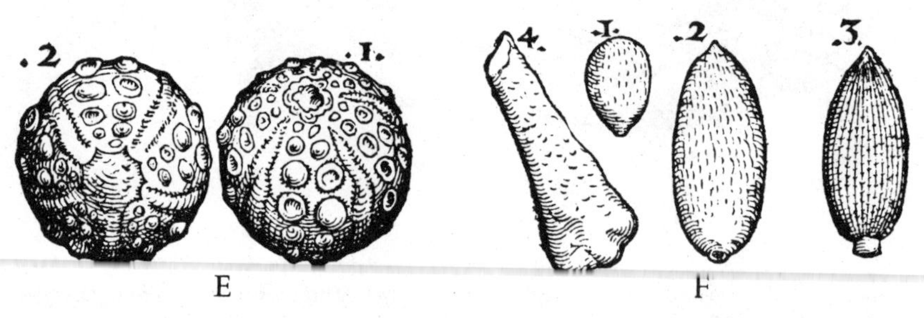

Fig. 1.10. *Gesner's woodcut illustrations (1558, 1565) of two species of living sea-urchins (A, B) and of three fossil echinoids (C, D, E) and their detached spines (F)[1.15]. He recognised a similarity between A and C, but the cidaroids D–F were unlike the 'regular' echinoids he knew alive (B: note the fine spines; the complex jaw-apparatus or 'Aristotle's lantern' has been exposed by dissection).*

preserved detached from the shell, were commonly considered to be acorn-like or fruit-like in shape, and Gesner placed them in the same chapter as his fossil wood; while for the tuberculated shells (Fig. 1.10 E) he fell back on a traditional notion that they were 'serpents' eggs', and assigned them to the chapter *On Stones which resemble Serpents and Insects*. In that chapter he also placed a loosely-spiralled ammonite, on account of its traditional interpretation as a coiled snake; whereas he noticed that a more tightly spiralled ammonite resembled ordinary gastropod shells (Fig. 1.7). Considering the difficulties already mentioned, it is hardly surprising that he saw no organic resemblance whatever in belemnite guards or crinoid ossicles: the dart-like belemnites and some wheel-like ossicles were assigned to the class of objects resembling human artefacts, while some star-shaped ossicles were placed in the class resembling heavenly bodies (Fig. 1.8).

In Gesner's work, then, we can see that even for one of the most competent naturalists of his age, with a remarkably wide and detailed knowledge of living animals and plants, the perception of resemblances between fossils and living organisms was far from straightforward. The fossil material in this collection was such that the resemblances he saw ranged from the obvious to the obscure. In terms of a spectrum it could not be simple to draw a line separating objects with such resemblances from those without.

IX

Gesner refrained from any explicit discussion of the causal origin of his 'fossils', partly because his book was only a preliminary survey of the subject. It may be that he suspected that some of his objects had organic resemblances simply because they were the genuine remains of organisms. His treatment of *glossopetrae* in the context of a description of living sharks suggests that he may have believed they were true sharks' teeth, and he may have come to a similar conclusion about some of his fossil shells, bones and wood. However this is likely to have seemed to him to be an explanation of peripheral interest and significance. For the whole framework of his classification shows that the central problem was that of resemblances of *any* kind. Why should many stones be formed in distinctive shapes at all, and why should so many of these shapes be reminiscent, not only of organisms, but of other entities in the universe?

Since Gesner was sufficiently sympathetic with Neoplatonic thought to mould his classification of 'fossils' around it, he probably found it congenial enough to regard many (if not all) of the resemblances he perceived as the manifestations of the hidden bonds of analogy and correspondence that linked the cosmos into a unity. Possibly he regarded some of the resemblances as merely fortuitous, and the resultant assignments to his classes as no more than convenient aids to identification. But it is much more likely, since he said his scheme 'followed the very steps and arrangement of nature', that he believed the resemblances were significant. Even a *trochite* (a circular crinoid ossicle—see Fig. 1.8, B) could resemble a man-made wheel, on this view, because both embodied in their different spheres the same supra-material Platonic *idea*. Likewise an *astroite* (a pentagonal crinoid ossicle—see Fig. 1.8, C) could be star-shaped because it owed its form to Hermetic stellar influence. Within the framework of Hermetic Neoplatonism none of the resemblances that Gesner noted had to be regarded as fortuitous. So also the stones that resembled animals and plants could owe those resemblances to their bonds of affinity with various organisms, and not to their origin as the remains of those organisms.

In this way the Neoplatonic thought of the sixteenth century made the modern interpretation of fossils less persuasive than it might otherwise have been, simply because it provided an alternative explanation of the resemblances between some 'fossil objects' and living organisms. There were inherent difficulties, as we have seen, in recognising such resemblances at all; but even when they were recognised, it did not seem to follow that those fossils must have been the remains of once-living organisms. That may seem to us to be an obvious inference to make, but to sixteenth-century naturalists it was far from convincing. Owing to the puzzles inherent in their material, most of the organic resemblances they could perceive at all seemed far from perfect, and it is not surprising that they often referred to them merely as 'images' (*imagines*) or 'pictures' (*icones*). If the cosmos was indeed a web of hidden affinities, it would be natural that many 'fossils' should imitate the forms of other entities. As for the causal origin of these forms, that could be attributed to the action of the same 'moulding force' (*vis plastica*) that governed the growth of living organisms, operating instead within the Earth.

Yet even where the natural philosophy of Neoplatonism was rejected or modified, the modern interpretation of the organic resemblances of fossils was still not persuasive. The reformed and purified Aristotelianism associated especially with the University of Padua was responsible for much fine biological research in the century that stretches from Vesalius's work to William Harvey's training there. On the problem of fossils, however, this Aristotelian thought, like the Neoplatonism it often opposed, had an interpretation of organic resemblances that seemed quite as persuasive as the modern explanation. Between living and non-living a sharper line was drawn than in Neoplatonic thought, and even the simplest organisms possessed a 'vegetative spirit' (*anima vegetativa*) that was manifest in their basic vital activities. However, it did not seem inconceivable that some at least of these activities could occur within the Earth. Simpler organisms were believed to be formed, at least on occasion, by 'spontaneous generation' (*generatio aequivoca*) from non-living material, and so their characteristic specific forms might be able to develop not only on the Earth's surface or in the sea, but also within the Earth, growing in this case from the 'stony' materials available there. More complex organisms might also be able to grow within the Earth, if their characteristic 'seed', holding their potential specific form, happened to penetrate there in percolating ground-water. For example, if the specific 'seed' of a fish were washed into the ground it might be able to grow from 'stony' material and generate a fish-*like* fossil in the rock. Such a fossil would have grown by a process directly comparable to the growth of a living fish, and it would owe its specific (say) perch-like form to an identical formative 'seed', but its matter would be different and it would *not* be the remains of a fish that had once been alive. It was plausible to explain in this way even those fossils that resembled marine organisms, since it was generally believed (mainly from observation of copious perennial springs) that there must be a direct subterranean circulation from the oceans, by which the 'seeds' of marine organisms could have been lodged in the Earth.

Thus in both Neoplatonic and Aristotelian terms it seemed rationally intelligible that among the whole range of 'fossils' there should be some with forms resembling living organisms. These might be produced by a moulding force that manifested the network of hidden affinities within the universe; or they might be generated by a process similar, in a limited way, to the generation of living organisms. In neither case, however, was it necessary to invoke the hypothesis that they were actually the remains of once-living animals and plants.

X

Both Aristotelianism and Neoplatonism thus provided satisfying interpretations of organic resemblances, and particularly of the fact that they varied from the obvious to the obscure. There was, however, a further reason why the modern explanation of these resemblances remained a hypothesis of peripheral importance to sixteenth-century naturalists. Even where the resemblances were most obvious—where the objects were at the 'easiest' end of the spectrum—there was a serious difficulty about accepting them as genuinely organic in origin. To make this inference involved accounting also for the changes in physical geography that were implied by the *position* in which the fossils were found.

This difficulty was minimal, once more, for geologically recent marine fossils, since these are commonly found in unconsolidated sediments on low ground and often close to the sea. It was fairly easy to conceive how changes in geography could have displaced land and sea and brought such fossils to the positions in which they were found. Ancient harbours were known to have been left miles inland by silting, and earthquakes were known to have occasionally the same effect. But for most of the commonest kinds of fossils such an explanation stretched the sober imagination too far, for they were often found on the tops of hills far from the sea. To assert that they were organic in origin therefore involved a belief in geographical changes far more radical than those for which there was good historical evidence. Moreover, many of these fossils were found embedded in solid rocks, which had no obvious resemblance to loose sediments; and it was therefore far from clear how the fossils, even if organic, could have got *inside* the rock.

There were two alternative solutions to this problem, if the fossils being discussed resembled living marine animals so closely that their organic origin seemed unavoidable. The first solution was to argue that they owed their position on hilltops to the action of Noah's Flood, which had risen high enough to cover even the highest hills. Tertullian and other patristic authors had used the existence of marine shells on hilltops as an argument for the universality of the Flood, in order to confute those contemporary pagan writers who asserted that the Flood had been a merely local inundation. This became a standard explanation in mediaeval writing. It always required some degree of flexibility in textual interpretation, since the Flood recorded in Genesis, if taken literally, had hardly been long enough for marine organisms to have migrated to the flooded areas, nor violent enough to have swept their shells up to the positions in which they were found. The literal meaning of Scripture, however, was the least important of the varied methods of exegesis that were sanctioned by the authority of Origen, Jerome and Augustine, and literalism was no essential ingredient of orthodoxy.

The literal meaning, however, while less edifying than allegorical interpretations, was the necessary foundation for them; and the efforts of Renaissance humanist scholars to recover the pure original text of Scripture, reinforced by the Protestant emphasis on its centrality for faith, focussed attention more closely on the plain meaning of narratives like that of Noah's Flood. To understand such passages correctly, commentators turned increasingly to the best secular knowledge of their day for aid and enlightenment. But this only heightened the problems they found they faced[37]. Not only did the increased range of biological knowledge pose problems about the recorded dimensions of the Ark, but more seriously a literally universal Flood required the production and subsequent disappearance of a huge volume of water, This could be explained, it was true, simply by attributing it to a miracle, but for many writers this seemed an unsatisfactory solution, for it placed the event outside the realm of natural law. Moreover, even if the water problem could be solved, a literally interpreted Flood would still be unsatisfactory as an explanation of marine fossils on hilltops.

The only alternative was to turn the patristic arguments upside down, and argue that pagan flood traditions were merely imperfect records of the Flood recorded in Genesis; that Deucalion for example was none other than Noah himself; and that the purpose of the Flood would have been adequately fulfilled if it was confined to the small areas then inhabited by Man. Although this solution raised further problems, such as why the building of an Ark had been necessary at all, it had the great virtue, in sixteenth-century eyes, of reconciling the Flood narrative to the natural philosophy of Aristotle. Noah's Flood became merely one of many local inundations that were natural to the globe. Aristotle's work *On Meteorology*, more geological in content than its title suggests, had outlined the continuous and generally gradual action of erosion and silting, which in the course of time could readily produce even major changes in physical geography. This was however integrated into Aristotle's eternalistic cosmos: "it is clear", he had concluded, "that as time is infinite and the universe eternal, that neither Tanais nor Nile always flowed ... for their action has an end whereas time has none."[38] Aristotle's eternalism was no stumbling-block to Christian thinkers merely on account of its extended time-scale, for a metaphorical exegesis of the 'days' of creation had been acceptable at least since the time of Augustine; but it did threaten the Christian doctrine of the createdness of the universe, by seeming to imply that God could not be fully transcendent over his creation.

The renewed interest in the literal exegesis of Scripture tended in the sixteenth century to heighten the problems of accepting an Aristotelian concept of the Earth's past history, because the calculations of historians and chronologers proceeded increasingly on the assumption that all the time-intervals used in Scripture, like those in secular chronicles, were strictly literal in connotation. Occasionally, however, it was possible to accept an Aristotelian picture of an ever-changing world of indefinite duration, for the purposes of natural philosophy, while at the same time to reject the metaphysical and theological implications of Aristotelian eternalism. This kind of compromise was encouraged particularly by the organisation of teaching at Padua, at which Aristotle's work was studied within a faculty of Arts and in a context primarily of medical education, and not within a faculty of Theology. Thus among natural philosophers and naturalists influenced by the Paduan tradition, if not among biblical commentators, it was not uncommon to accept an Aristotelian view of the ever-changing geography of the globe, with occasional local inundations of purely natural causation, of which the scriptural Flood had been but one. On this view, there might seem to be no problem at all about accepting the organic origin of fossils found on hilltops or even those embedded in layers of sediment. In practice however, this was not so straightforward, since it was difficult to *imagine* the radical changes in geography that were demanded by this explanation.

XI

Seen in retrospect, a modern interpretation of the organic resemblances of fossils was thus delayed by the lack of any satisfactory explanation of geographical change. At the same time its acceptance was also made less pressing, as we have seen, by the existence of two alternative interpretations of why some 'fossils' should resemble animals and plants. These alternatives seemed far more plausible as explanations of most of the commonest kinds of 'fossils'. Only those at the 'easiest' end of the spectrum resembled living organisms closely enough for the modern interpretation to be at all persuasive.

It is therefore not surprising that most of the early writers who have often been portrayed as championing a 'correct' interpretation of fossils are those who were referring to the 'easiest' kinds of fossils. These were generally the shells of marine molluscs of geologically recent date. Such fossils were well preserved; they belonged to extant groups and often even extant species; and they were usually found in unconsolidated sediments on low ground near the sea. The problems of matter, of form and of position were thus minimal. Isolated comments on the organic origin of fossils can be found in many writers back into classical Antiquity, but whenever localities are mentioned it is clear that they were referring to these fossils at the 'easiest' end of the spectrum. Even the earliest such writer, Xenophanes of Colophon (6th century B.C.), is said to have mentioned Malta and Syracuse, both of which are localities in which finely preserved Cainozoic mollusc shells can be collected in abundance.

Likewise it is well known that Leonardo da Vinci (1452–1519), more than half a century before Gesner wrote his book on 'fossils', recorded in his unpublished notebooks his reasons for believing that fossil shells were organic in origin.[39] His comments however, show that he was referring chiefly to fossils from the Cainozoic strata of northern Italy, which contain an abundance of beautifully preserved shells similar in general appearance to the shells of molluscs now living in the Mediterranean. Leonardo's notes certainly show much acute observation of living molluscs and their ecology, and of the processes of sedimentation. He recognised that the similarities between the fossils and living molluscs were so precise that a causal explanation was almost inescapable. He noticed that the fossil shells were similar not only in general form but also in many incidental features: for example they were preserved in various growth stages, and sometimes with other organisms adherent on them or bored into them. The occurrence of these fossils *within* a sediment was also no problem to him, since he understood enough of the process of silting and sedimentation to recognise the meaning of strata, and was able to account for their consolidation in terms of a process of 'drying out'.

This strikingly 'modern' interpretation, however, was relatively easy for Leonardo to make, not only because he was dealing with some of the 'easiest' kinds of fossils, but also because he had what he believed to be a satisfactory explanation of the geographical changes that his conclusion entailed. He was able to borrow from the mediaeval Aristotelian Albertus Magnus an explanation of how major interchanges of land and sea could have occurred without affecting the essential stability of the globe. Although he rejected the theory of spontaneous generation as an explanation of his fossils, Leonardo was favourably disposed in general towards the idea of stellar influences, and might indeed be regarded more truly as a Hermetic 'magician' than as a premature modern man of science. He also attacked the use of the Flood as an explanation of fossils, but this was not, as an earlier historical tradition supposed, an assault by an enlightened scientist on the prejudices of religious bigots. Leonardo was specifically attacking "ignoramuses", that is to say the unlearned; his essential motive was not to attack Christian orthodoxy but to defend the Aristotelian belief in the rational causality of natural events, which was difficult to apply to a literally interpreted Flood.

The same motive can be seen in the work of the Aristotelian physician Girolamo Fracastoro (1478?–1553). In order to explain 'action at a distance', such as magnetic attraction and infectious disease, without using the 'occult' Neoplatonic concepts of sympathy and antipathy, Fracastoro developed an earlier suggestion that *effluvia* or *seminaria* were being released continuously by such bodies as a lodestone or a diseased person (this has a spurious similarity to the much later germ theory of disease). However vague this explanation might seem, it had the virtue of bringing these otherwise mysterious phenomena within the scope of Aristotelian natural law. In 1517 fossil shells and crabs were discovered in foundations at Verona, and Fracastoro is said to have asserted that they were the true remains of shellfish, and to have ridiculed suggestions that they were due either to the Flood or to a moulding force within the Earth[40]. These comments show the same motives as the work for which he is more famous. To interpret the fossils as organic remains, and to attribute their emplacement to the ever-changing positions of land and sea, avoided *both* the miraculous overtones of the vulgar belief in a universal Flood, *and* the occult implications of the Neoplatonic explanation: the fossils were thus explicable in terms of natural law. From Fracastoro's Aristotelian point of view however, spontaneous generation was equally acceptable as an explanation, and in fact he is said to have used this for the interpretation of some more 'difficult' fossils.

Even for an Aristotelian, then, the organic explanation was not obligatory for all fossils. Falloppio, for example, as one of the great line of Paduan anatomists, would have been well aware of organic resemblances in whatever fossils he was familiar with; yet he felt that their occurrence on hilltops far from the sea made their organic origin unacceptable. It would have entailed changes in geography that were literally incredible, even with an Aristotelian view of continuous terrestrial change; and an explanation in terms of spontaneous generation therefore seemed preferable.

For Neoplatonists too, the acceptance of an organic origin for some of the 'easier' fossils did not entail the same explanation for more 'difficult' specimens. Thus Cardano repeated arguments similar to Leonardo's—just possibly by knowledge of the unpublished notebooks—and clearly believed that some fossil shells betrayed changes in the position of land and sea[41]. He did not however ascribe these changes to a single Flood, for he believed there had been many local inundations. Yet most of the fossils he described were attributed in Neoplatonic terms to the action of a moulding force, and Cardano believed that their characteristic forms possessed some Hermetic power.

It is interesting to see how Cardano's organic interpretation of his 'easier' fossils could be misconstrued by a reader who was unfamiliar with its Aristotelian background. Palissy misread Cardano (in a French translation) and assumed that he had attributed these fossils to a universal Flood. Palissy himself wished to assert the organic explanation, for he was well aware of the similarities between many fossil shells and living shellfish, and at the same time found other explanations of those similarities unacceptable. Yet he rejected the vulgar notion of a universal Flood as inadequate to explain the widespread occurrence of the fossils far above the present sea-level[42]. He was therefore in a dilemma which he was only able to resolve by arguing that the inland fossil shells, although truly organic, were not marine but freshwater in origin. But this solution then faced him with a further problem, since many of the fossils clearly resembled marine species, and furthermore he knew that they were much more diverse than the restricted range of living freshwater animals. Palissy could only meet this problem by suggesting awkwardly that some of the former inland lakes, in which the fossil shells had originated, had been rather salty and therefore able to support species of marine aspect, and that some of the more edible species had long since disappeared through over-fishing. This somewhat tortuous reasoning is a good illustration of the difficulties that any sixteenth-century naturalist faced in asserting the organic origin of fossils. On the other hand, the petrified substance of some of his fossils, and the solidity of the enclosing rocks, was no problem to Palissy, since he believed that percolating 'salts' could have affected this change with great rapidity.

Colonna may be taken as a final example of naturalists who asserted the organic origin of some fossils. His *Observations on some aquatic and terrestrial Animals* (1616) are important in more than one respect[43]. While many earlier naturalists, and supremely Gesner, had been familiar with a wide range of living organisms, they described their fossils, as we have seen, within an essentially mineralogical context. Colonna, on the other hand, was one of the first to place them instead in a primarily biological context, and to describe them alongside whatever living organisms they resembled (Fig. 1.11). This did not necessarily lead to an acceptance of their organic origin, and several excellent naturalists later in the seventeenth century continued to have well-founded doubts on this point. But it did serve to focus attention more closely on the precise nature of the resemblances; and it is no accident that, from Colonna onwards, most of those who did argue for the organic origin of fossils were primarily biologists. Colonna also applied the same precise nomenclature to his fossils as to his living animals, distinguishing different kinds of related fossils with more accuracy than ever before. He also grasped the relation between shells and the casts or moulds that they left in the fossil state, and thereby overcame the inherent problems of 'matter' sufficiently to recognise a wide range of fossils as organic in origin. In particular, he wrote a special essay on 'tongue-stones' (*glossopetrae*), arguing that they were the true teeth of

Fig. 1.11. One of Fabio Colonna's plates of copper engravings (1616)[9]. Note the careful differentiation of several living species of whelk shells ('Buccinum'), and the inclusion of a fossil specimen (top left) within the same scheme and on the same plate. This is one of the earliest uses of copper engraving for the illustration of fossils: note the economical use of the space available.

sharks, and he pointed out that they often occurred with the shells of oysters and other marine molluscs[44]. These conclusions, however, still faced him, as they faced any other naturalist, with the problem of the position in which the fossils were found; and although many of Colonna's fossils came from the hills of Puglia and were embedded in solid strata, he saw no alternative but to attribute them to the Flood.

XII

Judged from the standpoint of modern palaeontology it might seem that this analysis of sixteenth-century studies on fossils is a story of failure. It is true that Gesner's work incorporated important innovations for the future development of the *activity* of a science of palaeontology, yet neither he nor his contemporaries made more than marginal progress towards recognising the organic origin of the fossils with which they were familiar.

It is questionable, however, whether we *should* make such judgements. From a historical point of view such apparent 'failures' may be more revealing than the more obvious 'success stories' of science. The 'failure' of a fine naturalist like Gesner to arrive at a clear conclusion about the origin of fossils, despite his wide knowledge of living animals and plants, may tell us more about the world of sixteenth-century science than the isolated 'correct' remarks about fossils that were so assiduously collected in an earlier historical tradition. We have seen that the problem of fossils was not a simple one of deciding on their organic origin, but a complex matter of discriminating the organic from the inorganic within a continuous spectrum of 'fossil objects'. There were limitations, inherent in the nature of the material available and in the state of biological knowledge, on the range of objects in which organic resemblances were likely to be seen. However, even when such resemblances could be seen, there was the further problem of the position in which many fossils were found; and in the absence of satisfactory explanations of geographical change this was liable to rule out an organic interpretation. In general, therefore, the modern interpretation was only applied in cases where the problems of matter, of form and of position were all minimal; and this inevitably restricted the applicability of the organic explanation to a small proportion of all the fossils that were known.

Beyond all these difficulties lay a more serious intellectual problem. Even when resemblances between fossils and living organisms could be clearly perceived, it did not seem to follow necessarily that the fossils were actually the remains of living organisms. This inference, so obvious to us today, was not avoided in the sixteenth century for reasons of intellectual conservatism or out of any sense of conflict with religious orthodoxy. It was usually ignored or rejected on the far more positive grounds that it was not a necessary inference within *either* of the two dominant—and 'progressive'!—intellectual frameworks of the time. Both the renewed Aristotelianism and the synthetic Neoplatonism of the sixteenth century can be seen in retrospect to have contributed much to the later development of 'modern science'; but on the question of fossils both of these natural philosophies provided the phenomenon of organic resemblance with explanations that were quite as persuasive, ndeed more so, than the hypothesis of organic origin. Aristotelians could attribute organic resemblances to the growth *in situ* of objects combining the form of genuine organisms with the stony matter appropriate to all 'fossils'; objects for which the causal explanation lay in spontaneous generation or the implantation of specific 'seeds' within the Earth. Neoplatonists could attribute the same resemblances to the action of a pervasive moulding force or 'plastic virtue', which made visible the hidden web of affinities that bound all parts of the cosmos into one. In either case, the explanations successfully accounted for the fact that the resemblances varied from the striking to the barely perceptible, and they were therefore more widely applicable and more 'successful' than the hypothesis of organic origin.

With such powerful alternatives available, no single observation or specimen, however striking, could be decisive in favour of a wide-ranging theory of the organic origin of fossils. It might erode the edge of the established interpretations, by detaching certain objects and transferring them to the category of those for which an organic origin was acceptable; but it could scarcely undermine the explanatory power of the alternatives in accounting for the majority of 'fossil objects'. The organic explanation could not be extended to a broader range of objects until after the credibility of the alternative interpretations of organic resemblance had broken down; but that in turn required changes in the dominant philosophies of nature, extending far beyond the problem of fossils.

REFERENCES

1. Conradus Gesnerus, *De Rerum fossilium, Lapidum et Gemmarum maxime, figuris et similitudinibus Liber*, Tiguri, 1565. Willy Ley, 'Konrad Gesner, Leben und Werk', *Münchener Beiträge zur Geschichte und Literatur der Naturwissenschaften und Medezin*, Heft 15/16, München, 1929. Gerald P. R. Martin, 'Conrad Gesner. Zu seinem vierhundertsten Todestage am 13. Dezember 1965', *Natur und Museum*, Frankfurt, 1965, vol. 95, pp. 483–494.

2. M. J. S. Rudwick, 'Problems in the Recognition of Fossils as organic Remains', *Actes, Xme Congrès internationale de l'Histoire des Sciences* (1964), pp. 985–7.

3. Conradus Gesnerus, *Historiae Animalium*, Tiguri, 1551–8, 4 vols.

4. Ulyssus Aldrovandus, *Musaeum Metallicum in Libros IIII distributum . . .*, Bononiae, 1648. This was edited by Bartolomeo Ambrosino, probably from the material mentioned in Aldrovandi's will as '*Geologia* ovvero *De Fossilibus*' (see Adams, *Birth and Development*, p. 166).

5. Georgius Agricola, *De Natura Fossilium Lib. X*, Basiliae, 1546. For a modern English translation, see M. C. Bandy and J. A. Bandy, 'De Natura Fossilium (Textbook of Mineralogy) by Georgius Agricola', *Geological Society of America, Special Paper* no. 63 (1955).

6. Gesner, *De Rerum fossilium, Epistola dedicatoria.*

7. Leonardus Fuchsius, *De Historia Stirpium Commentarii*, Basiliae, 1542; Andreas Vesalius, *De Humani Corporis Fabrica Libri Septem*, Basiliae, 1543.

8. Christophorus Encelius, *De Re metallica, hoc est. de Origine, Varietate & Natura Corporum Metallicorum, Lapidum, Gemmarum, atq. aliarum, quae ex Fodinis cruuntur, Rerum, ad Medicinae Usum deseruientium, Libri III*, Francofurdi, 1557.

9. Fabius Columnis, *Aquatilium, et Terrestrium aliquot Animalium, aliarumq. naturalium Rerum observationes*, Romae, 1616.

10. C. E. Raven, *Natural Religion and Christian Theology;* first series, *Science and Religion*, Cambridge, 1953. See Chapter 5, 'Gesner and the Age of Transition'.

11. Io. Kentmanus, *Nomenclaturae Rerum fossilium, quae in Misnia praecipue & in aliis quoque regionibus inveniuntur*, Tiguri, 1565.

12. Michaelus Mercatus, *Metallotheca vaticana*, Romae, 1719 (although published so long after its compilation, it was known in manuscript and was used, for example, by Steno in the seventeenth century); Andreus Caesalpinus, *De Metallicis Libri tres*, Romae, 1583. J. B. Olivus, *De reconditis et praecipuis Collectaneis ab Francesco Calceolario veronensi in Musaeo adservatis*, Veronae, 1584. Benedictus Cerutus and Andreus Chioccus, *Musaeum Franc. Calceolari iun. Veronensis . . .*, Veronae, 1622.

13. Conradus Gesnerus, *De omni Rerum fossilium Genere, Gemmis, Lapidibus, Metallis et huiusmodi, Libri aliquot*, Tiguri, 1565–6.

14. H. M. Fisch, 'The Academy of the Investigators', in E. A. Underwood (ed.), *Science, Medicine and History*, Oxford, 1953, vol. 1, pp. 521–563.

15. Conradus Gesnerus, *Historiae Animalium Liber IIII, qui est de Piscium & aquatilium Animantium Natura*, Tiguri, 1558.

16. Gesner, *De Rerum fossilium, Epistola dedicatoria.*

17. Agricola, *De Natura Fossilium; De Re metallica*, Basiliae, 1556: for a modern English translation of the latter, see H. C. Hoover and L. H. Hoover, *Georgius Agricola. De Re Metallica*, London, 1912.

18. Allen G. Debus, *The Chemical Dream of the Renaissance*, Cambridge, 1968.

19. Bernard Palissy, *Discours Admirables, de la nature des eaux et fonteines, tant naturelles qu'artificielles, des metaux, des sels et salines, des pierres, des terres, du feu et des emaux, avec plusieurs autres secrets des choses naturelles, plus un traite de la marne fort utile et necessaire pour ceux qui se meslent de l'agriculture, le tout dresse par dialogues esquels sont introduits la Theorique et la Practique*, Paris, 1580. For a modern English translation, see A. La Rocque, *The Admirable Discourses of Bernard Palissy*, Urbana, 1957. See also A. La Rocque, 'Bernard Palissy', *in* Cecil J. Schneer (ed.), *Toward a History of Geology*, Cambridge (Mass.), 1969, pp. 226–241.

20. Bernard Palissy, *Recepte véritable par laquelle tous les hommes de la France pourront apprendre a multiplier et augmenter leurs thresors, Item, ceux qui n'ont jamais eu cognoissance des lettres pourront apprendre une philosophie necessaire a tous les habitants de la terre . . .*, La Rochelle, 1563.

21. Agnes Arber, *Herbals: their Origin and Evolution. A Chapter in the History of Botany, 1470–1670*, Cambridge, 1953.

22. [Conrad Gesner], *Thesaurus Euonymus Philatri, de Remediis secretis . . .*, Tiguri, 1554.

23. Gabrielus Falopius, *De medicatis Aquis, atque de Fossilibus, Tractatus*, Venetiis, 1564. See also J. Bauhinus, *Historia novi et admirabilis fontis balneique Bollensis in Ducatu Wirtembergio . . . Adijciuntur plurimae figurae novae variourum fossilium, stirpium & insectorum, quae in & circa hunc fontem reperiuntur*, Montisbeligardi, 1598.

24. Aldrovandus, *Musaeum Metallicum;* see Liber IV, *De Lapidibus in genere.*

25. A. G. Debus, *The English Paracelsians*, London, 1965, see Chapter I.

26. Frances E. Yates, *Giordano Bruno and the Hermetic Tradition*, London, 1964.

27. Frances E. Yates, 'The Hermetic Tradition in Renaissance Neoplatonism', *in* C. S. Singleton (ed.), *Art, Science and History in the Renaissance*, Baltimore, 1967.

28. Camillus Leonardus, *Speculum Lapidum*, Venetiis, 1502. L. Thorndike, *A History of Magic and experimental Science: the sixteenth Century*, New York, 1941.

29. Hieronymus Cardanus, *De Subtilitate Libri XXI*, Nuremberg, 1550 (and later editions).

30. Cardanus, *De Subtilitate:* see *Liber VII, De Lapidibus.*

31. Franciscus Rueus, *De Gemmis aliquot, iis praesertim quarum Divus Ionannes Apostolus in sua Apocalypsi meminit: de aliis quoque, quarum usus hoc aevi apud omnes percrebruit, Libri duo . . .*, Tiguri, 1565.

32. Epiphanius, *De XII Gemmis, quae erant in Veste Aaronis, Liber graecus . . . cum Corollario Conradi Gesneri*, Tiguri, 1566.

33. For a broader discussion of such apparently 'miscellaneous' compilations, see Michel Foucault, *Les Mots et les Choses. Une Archéologie des Sciences humaines*, Paris, 1966 (English translation, *The Order of Things*, London, 1970), ch. 2.

34. Ioannus Kentmanus, *Calculorum qui in Corpore ac Membris Hominum Innascuntur, genera XII depicta descriptaq., cum Historiis singulorum admirandis*, Tiguri, 1565.

35. See Allen G. Debus, 'Edward Jorden and the Fermentation of the Metals: An Iatrochemical Study of Terrestrial Phenomena', *in* Schneer, *Toward a History of Geology*, pp. 100–121.

36. Guilielmus Rondeletius, *Libri de Piscibus marinis, in quibus verae piscium effigies expressae sunt*, Lugduni, 1554; Pierre Belon, *La Nature et Diversité des Poissons*, Paris, 1555.

37. D. C. Allen, 'The Legend of Noah, Renaissance Rationalism in Art, Science and Letters', *University of Illinois Studies in Language and Literature*, vol. 33, nos. 3–4, Urbana, 1949 (reprinted 1963). See also John Dillenberger, *Protestant Thought and Natural Science. A Historical Interpretation*, London, 1961, chs. 1–3.

38. Aristotle, *Meteorologica* (transl. H. D. P. Lee), London, 1952, book I, ch. 14.

39. Edward MacCurdy, *The Notebooks of Leonardo da Vinci*, London, 1938: see vol. I, pp. 325–374. The main passages come from a manuscript dated about 1508–9.

40. Cerutus and Chioccus, *Musaeum Franc. Calceolari*, p. 407ff.

41. Cardanus, *De Subtilitate*, see *Liber II, De Elementis & eorum Motibus & Actionibus*, and *Liber VII, De Lapidibus.*

42. Palissy, *Discours admirables;* see also H. R. Thompson, 'The geographical and geological Observations of Bernard Palissy the Potter', *Annals of Science*, vol. 10, pp. 149–165, 1954.

43. Columnus, *Observationes:* see Cap. XXI, *De varia lapidum concretione, & rebus in lapidem versis eorum effigie remanente.*

44. Fabius Columnus, *De Glossopetris Dissertatio*. In *Fabii Columnae Lyncei Purpura*, Romae, 1616, pp. 31–39.

Chapter Two

Natural Antiquities

I

ONE day in October 1666, some fishermen brought a huge shark ashore near Livorno. It was one of those occasional chance events that have far-reaching consequences for the history of science. Falling, as it were, on intellectually prepared ground, it had a striking catalytic effect: it led to the dissolution of the stable situation described in the last chapter, and to the introduction of a new dimension into the debate about fossils.

The shark was landed within the realms of the Grand Duke of Tuscany, Ferdinand II, who, as a generous patron of the sciences, ordered its head to be taken to Florence to be dissected by the anatomist Niels Stensen (1638–1686)[1]. Stensen, who is better known by his literary name Stenonis (anglicised to Steno), had left his native Copenhagen in 1660 to pursue his medical studies at Leiden, which by this period had won the pre-eminence as a medical centre that Padua had had in the previous century. Because no university post was available for him in Denmark, Steno had then moved on, working briefly at Paris and Montpellier, and had arrived in Florence in 1665—exactly a century after Gesner's death. The quality of his anatomical work had already been recognised, and he was appointed by Duke Ferdinand to a hospital post that provided him with a living and allowed him ample time for research. He was also elected to the Experimental Academy (*Accademia del Cimento*), which had been founded at Florence by the Duke's brother, Leopold de' Medici, with ideals similar to those of the earlier Academy of the Lynxes at Rome. More specifically, the Experimental Academy sought to perpetuate and extend the experimental and mathematical approach to science that Galileo had used with such success earlier in the century.

It is therefore not surprising that Steno's research, after his arrival in Florence, turned first towards the application of similar methods to the distinctively biological problem of muscular contraction, on which he had begun to work while at Leiden. By analysing the nature of contraction in geometrical terms, he was able to confirm that the apparent swelling of a contracted muscle was simply due to the shortening of the fibres without change of volume, and not, as was commonly believed, to a true swelling of the muscle. He was working on this "geometrical system of the muscles" when the shark's head arrived for dissection. Having studied briefly the perishable organs of the head, he was able to examine the teeth at greater leisure, and it was this that led him to consider the old problem of the nature of fossil 'tongue-stones'. He had probably been familiar with such objects ever since his early university training in Denmark, for an illustrated catalogue of the celebrated collection of Ole Worm (1588–1654) in Copenhagen had just been published, and this contained many 'fossil objects' among other natural history specimens and antiquities[2]. Probably Steno also found ample material in Duke Ferdinand's collections; but it seems to have been his examination of the teeth of the giant shark that first diverted his attention on to the problem of fossils, by leading him to believe that a convincing case could be made for the organic origin of tongue-stones. He appended his arguments on this point as a long "digression" at the end of his study of *The head of a shark dissected*, which in turn he published in the same volume as his *Sample of the elements of myology, or, A geometrical description of muscles* (1667)[3].

In order to demonstrate that tongue-stones were the true teeth of fossil sharks, Steno had to show first that there was no evidence for their growth *in situ* within the rocks; and he asserted that on the contrary they often showed signs of decay, implying that they were not being formed at the present time but were relics of an earlier period. He then argued that the surrounding 'earth' must have been soft when it first enclosed them, for they were not distorted like the roots of trees that grow within the crevices of rocks. The original softness of the 'earth' he ascribed to its mixture with water, either at the time of the Creation or at the Flood, and he argued that its layered appearance must be due to gradual settling out or precipitation of sediment.

"How well then everything agrees", he exclaimed, "how unanimously they all point in the same direction". He now saw no obstacle to the conclusion that tongue-stones were the teeth of sharks that had died during the period of sedimentation. This was further supported by their manifest identity of *form*, which he illustrated by printing engravings from Mercati's still unpublished catalogue of the Vatican collections (Fig. 2.1); indeed he felt it was a striking fact that although tongue-stones were such complex objects they showed fewer defects from their 'ideal' form than much simpler entities such as crystals. Furthermore, their difference of *matter* when compared with the teeth of living sharks, whether stony of friable, could be simply explained, he believed, as the result of either impregnation or the loss of volatiles; and the elevated *position* in which tongue-stones were found, for example in the famous localities in Malta, could be attributed to a subterranean upheaval of some kind. For these admittedly 'easy' fossils Steno had adequate solutions for the inherent problems of form, matter and position, and was therefore able to argue for their organic origin with some confidence.

The context of Steno's essay on tongue-stones is similar to that of Colonna half a century earlier, although Steno did not cite Colonna's work and may not have known of it. Like Colonna, Steno came to the problem of fossils from primarily biological interests, and was well placed to appreciate the detailed resemblances between tongue-stones and the teeth of living sharks. However, Steno's essay also has several features which reflect the Galilean tradition in which he was working, and which contrast strongly with most previous treatments of the subject, if not with Colonna's. Firstly, the problem was considered solely as one of efficient causation: the purposes of tongue-stones, and specifically their medical or natural-magical 'virtues', were not mentioned even to be refuted. This narrowing of the motives for interest in fossils was to become characteristic of the debate during the rest of the century. Secondly, Steno abandoned the earlier encyclopaedic tradition of compiling all previous opinions on the subject, and scarcely mentioned any but his own contemporaries: this perhaps reflects the growing confidence of the 'moderns' in their ability to surpass the achievements of the 'ancients' in every field. Thirdly, he made a sharp distinction between his observed facts about tongue-stones and his "conjectures" from these facts, listing the observations first and then citing them at appropriate points in his argument, almost in the manner

Fig. 2.1. *Steno's illustration (an earlier but unpublished plate by Mercati) of a shark's head and of its teeth (1667)*[3]. *He compared the teeth with the much larger fossil 'tongue-stones' (Glossopetrae), and argued in detail for a causal explanation of the similarity, that is that the Glossopetrae were fossil teeth of very large sharks.*

of a mathematical theorem. Although this mathematical form was more apparent than real, he was certainly trying to construct a connected persuasive argument, leading step by step from observation through inference to conclusion. This reflects an awareness of the problems of method in science, and of the need to conduct scientific discussion within a community committed to rational argument on the basis of empirical observation. Thus, although Steno was probably convinced in his own mind that tongue-stones were the true teeth of fossil sharks, he disowned any claim to certainty in the matter: he said that his essay would merely present the case for their organic origin, which could then be countered, as in a law-suit, by the opposite case for their origin *in situ* within the rocks.

II

Steno's comments on fossils, although embedded in a primarily anatomical work, did not remain unnoticed. Indeed, his work was brought to the attention of English naturalists within a few months of its publication, by the new medium of the scientific periodical. The secretary of the Royal Society, Henry Oldenburg, included an abstract of *The head of a shark dissected* in one of the early issues of his news-sheet for the Fellows of the Society, the *Philosophical Transactions*[4]. However, he could not refrain from commenting that the infant Royal Society was in no way lagging behind the Italian scene in the attention it had already given to the puzzling problems of fossil objects. The Society's 'Curator of Experiments' Robert Hooke (1635–1703) had, he pointed out, already lectured on the subject. Hooke may have met Steno in 1665, while both of them were at Montpellier, and it is even possible that Steno derived some of his ideas about fossils from this contact[5].

The achievements of the Royal Society in the realm of physical science have tended to obscure the wide range of interests of its Fellows, or, more seriously, to suggest that outside their mathematically inclined research they were merely indulging in sterile fact-collecting or dilettante dabbling in science. An indication of the inadequacy of that view is the fact that on one occasion when Hooke wished to reassure the Society that its endeavours were not futile, he actually chose the problem of fossils as a paradigm example of the fruits of knowledge to be gained by following the 'experimental' method of enquiry[6]. Although this problem was not at all susceptible of mathematical treatment, he believed it would nevertheless serve to show a sceptical and often hostile intellectual world that the patient collection of observations and careful reasoning upon them was indeed capable of yielding positive knowledge. Likewise, although his book *Micrographia* (1665) was ostensibly designed to illustrate the new dimension of natural knowledge that was opened up by the use of the microscope, Hooke was concerned more fundamentally to use this novel extension of the senses as a vindication of the Society's philosophical programme; and it was therefore not inappropriate to use a microscopical observation on fossil wood as a starting-point for a brief general essay on the origin of fossils.[7]

Hooke observed that the micro-structure of some specimens of fossil wood resembled closely that of rotten or charred pieces of ordinary wood (Fig. 2.2). He attributed the stony matter of the fossil wood to its "having lain in some place where it was well soak'd with *petrifying* water" which, he argued, had impregnated it with "stony and earthy particles". He then extended this explanation to cover other stony objects with organic resemblances. In particular he successfully sorted out the confusing modes of preservation of ammonites, and recognised their general similarity to the chambered shells of the pearly nautilus (which was known by now from explorations in the East Indies, though still a great rarity). That no closer living analogues were known does not seem to have disturbed him at this stage, and he attributed the position of his fossils on land simply to "some Deluge, Inundation, Earthquake, or some such other means".

But although he thus surmounted the problems of form, matter and position, Hooke's most persuasive reason for believing in the organic origin of these fossils was derived from the philosophical principle of sufficient reason: any other explanation, he argued would have been "quite contrary to the infinite prudence of Nature", that Nature which, in the ancient phrase, "*does nothing in vain*"[8]. As long as the form of 'fossil objects' was believed to reflect their cosmic affinities and therefore to express their medicinal or magical value, their varied resemblances could be assimilated to a teleological view of the natural world. However, once that belief became unacceptable, as it was to Hooke, the resemblance between a fossil and an organism became inexplicable, unless it had a straightforward causal origin. A fossil shell with a resemblance to the functional shell of a living mollusc could not be brought within a designful universe unless it too had served to

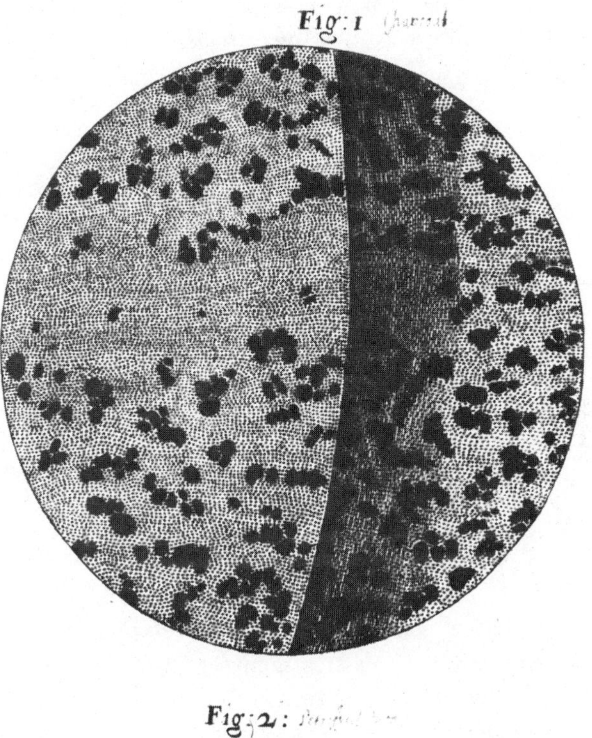

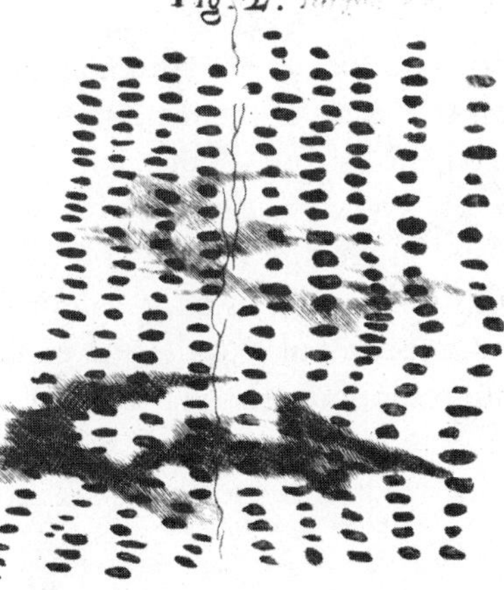

Fig. 2.2. Robert Hooke's comparison (1665)[7] *between the microstructure of charcoal* (above) *and that of fossil wood* (below)*: this was probably the first time the microscope was used to throw light on the problem of the origin of fossils.*

protect a living animal. The teleological view, grounded in the natural theology of the time, was in no way rejected: it was simply diverted into a powerful argument in favour of the organic origin of fossils.

III

The explanation that Hooke opposed was, in his own words, the view that fossils owed "their formation and figuration" to some "kind of *Plastick virtue* inherent in the earth". The continuing popularity of this view, stemming as we have seen from the Neoplatonism of the previous century, owed much to the work of one of the most prolific and versatile scholars of the age, the German Jesuit Athanasius Kircher (1602–1680). Kircher's highly popular encyclopaedia on *The Subterranean World* (1664) described the "geocosm" of a static Earth in terms of an extended organismic analogy with the microcosm[9]. The stony matter of 'fossil objects' was attributed to a "lapidifying virtue diffused through the whole body of the geocosm", and their form to a "*spiritus plasticus*" analogous to that which controlled the development of an organism. It may well have been this treatment of fossils that was in Hooke's mind when he wrote his comments on fossils for *Micrographia.* No stony resemblance or likeness was too implausible for Kircher to believe, and he decorated his work with a fantastic collection of supposed natural 'images', many of them clearly derived—with further imaginative 'improvement'—from Aldrovandi's already somewhat idealised illustrations. It would be easy to dismiss Kircher's work as merely 'unscientific' and unworthy of serious attention; but for all its credulity it was highly influential in expressing an interpretation of 'fossils' that continued to be popular to the end of the century. Within its own philosophical presuppositions such an interpretation continued to be quite satisfactory as an explanation of the whole varied spectrum of fossil objects; and where it was rejected, it was for primarily philosophical reasons.

This can be seen particularly clearly in a book on fossils by the Sicilian painter and naturalist Agostino Scilla (1639–1700), published only three years after Steno's essay (of which Scilla was apparently unaware). The essential argument of the book is epitomised in its title and symbolic frontispiece (Fig. 2.3) of *Vain Speculation undeceived by Sense* (1670)[10]. A solid figure of 'Sense' (or better, 'Sense-experience')

Fig. 2.3. The frontispiece of Scilla's book on fossils (1670)[10]*, showing 'Sense' with the eye of Reason demonstrating to 'Vain Speculation' the organic nature of a fossil sea-urchin and shark's tooth. Note the other fossils strewn on the ground: Scilla explained the origin and position of fossils in terms of a general flood.*

was shown demonstrating the manifestly organic nature of a tongue-stone and a fossil echinoid to a wraithlike 'Speculation'. The latter figure represents not, as an earlier historical tradition tended to assume, the benighted opinions of the Church, but the natural philosophy that attributed fossils to growth *in situ* in the rocks. For Scilla, as for Steno, it was this opinion that required refutation; and indeed Kircher may have been the unnamed adversary against which the book was directed. However, it is clear that he saw the problem primarily as a clash between conflicting philosophical positions. Like Steno, however, he was working with relatively 'easy' fossils, mostly Cainozoic faunas from Malta, Calabria and the environs of his native Messina (Fig. 2.4). These were similar in form to living animals with which he was familiar, they were little altered in substance apart from some impregnation, and they were found quite close to present sea-level. Their organic origin was therefore fairly simple to assert.

Unlike Scilla, Steno seems to have realised that the refutation of views such as Kircher's required more than the demonstration of the organic origin of certain relatively 'easy' fossils. Within the whole spectrum of 'fossils', some objects (such as rock-crystal) fairly clearly *did* 'grow' within the Earth, while others (such as tongue-stones) were, he believed, extraneous. Some criteria were needed for distinguishing these categories from each other. After his preliminary essay had been published, Steno continued to struggle with this problem, enlarging his knowledge of the phenomena by first-hand fieldwork in Tuscany and elsewhere. He had begun to plan a full-scale work on the subject when he was summoned by the King of Denmark to return to a well-paid post in Copenhagen, and he had to content himself with publishing a brief *Forerunner* (1669) of his projected *Dissertation on a solid naturally enclosed within a solid*[11].

In modern terms the contents of Steno's *Forerunner* have seemed so diverse as to justify awarding him the title of 'founder' of such disparate sciences as crystallography, palaeontology and stratigraphy; but this judgment is anachronistic, for in the terms of its own time Steno's sketch of his projected work can be seen to form a unified argument. The title of the work, so strangely uninformative at first sight, reflects the wide context in which he now recognised that he had to tackle the problem of fossils. Characteristically he stated the central question in the form of a theorem: "Given an object possessing a certain form, and produced by natural means, to find in the object

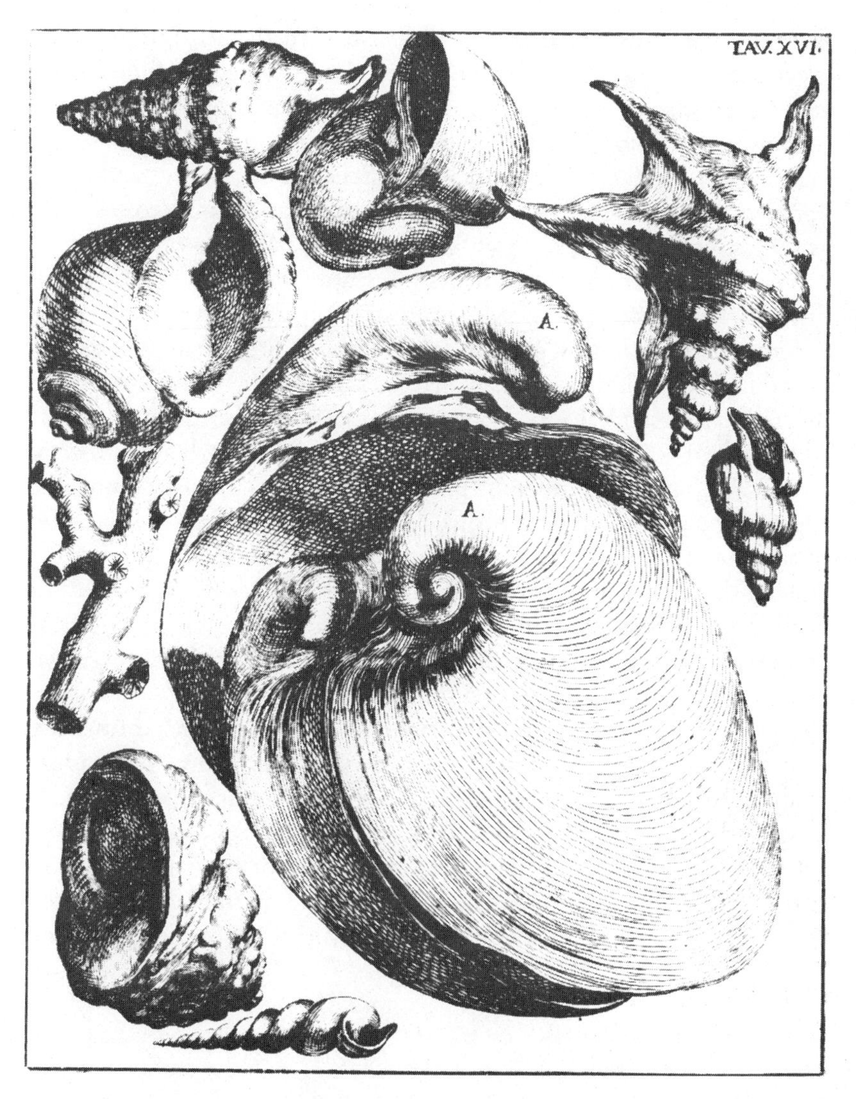

Fig. 2.4. Fossil shells and corals, illustrated in one of the plates from Scilla's book (1670)[10]. *Being well preserved and similar to living species, their organic origin could be accepted easily.*

itself evidence showing the position and manner of its production." The problem was that of determining the origin of *any* 'fossil' of distinctive form, whether it was a tongue-stone, a stony shell, or a crystal. In order to solve this problem, and to distinguish 'fossils' that were organic remains from those that had been formed *in situ* within the Earth, Steno saw that the crucial evidence would be found in an analysis of their modes of growth. For having rejected Kircher's organismic analogies for terrestrial features, he was well prepared to believe that inorganic 'fossils' such as crystals did not 'grow' within the Earth in the same sense or in the same manner as organic 'fossils' such as tongue-stones had grown within the bodies of living organisms.

Steno used a corpuscular theory of matter to make precisely this distinction. He analysed the varied forms of quartz and pyrite crystals in terms of accretionary growth from particles precipitated from surrounding fluids, and concluded that such naturally occurring crystals differed in no essential way from those produced experimentally in the laboratory. The varied forms of mollusc shells, on the other hand, were due to a significantly different pattern of accretion, chiefly along the edges of the shells, and clearly owed their growth to the vital activities of the animals they enclosed; and in this respect fossil shells were the same as those of living molluscs. From these analyses of the growth of 'fossil objects' Steno had in effect derived criteria by which to distinguish objects analogous to chemists' crystals, which could have grown *in situ* from percolating fluids within the Earth, from objects analogous to parts of living organisms, which could only have been formed by such organisms and could *not* have grown *in situ*. Moreover, having used percolating fluids to explain the growth of inorganic 'fossils', Steno was able to use them also to explain the residual differences between fossil and living shells: any differences of substance could be attributed either to impregnation with extra particles precipitated from such fluids or to the leaching out of some of the original particles.

IV

Before the *Forerunner* was published Steno had left Florence to return to Denmark; and although he later came back he made no further contributions to the problem of fossils, and the full *Dissertation* was never written. As Leibniz and other contemporaries commented with regret, Steno's conversion to Catholicism and subsequent ordination seem to have diverted his attention wholly away from any interest in natural science. His work, however, did not suffer from neglect. The *Forerunner* was published in Florence and in Amsterdam, and on reaching

England was translated by Oldenburg. However, it seems to have been valued by the Royal Society circle as much for its illustration of corpuscularian matter-theory as for its conclusions about fossils: Oldenburg in his introduction suggested that it was as important for its analysis of 'earth' as Robert Boyle's recent work on air[12]. Although Steno had deliberately avoided entangling his argument in the current disputes about the nature of matter, he was explicitly committed to a general corpuscular theory, and it was this that had enabled him to distinguish so clearly between the growth of crystals and the growth of shells. Moreover it was the Florentine climate of 'mechanical philosophy' that had set him looking for an explanation of the efficient cause of the form of fossils, and had made him dissatisfied with any assertion that fossils were 'produced by nature' unless the agency was specified. His work was therefore appreciated in the Royal Society not least as an example of the 'new philosophy' applied to the old problem of the 'growth of stones'. Boyle, for instance, felt it relevant to re-issue Oldenburg's translation of the *Forerunner* with his own *Essay about the origin of Gems* (1672)[13]: while sceptical about many of the alleged natural-magical powers of stones, he believed that such powers as were genuine could be explained within the mechanical philosophy, if stones were taken to have a corpuscular constitution.

It would be misleading to regard Kircher's work as typifying the opposition to Hooke's and Steno's organic theory of fossils. Kircher had indeed studied some phenomena of *The subterranean World* at first hand—his descriptions of volcanoes are particularly good—but he was no biologist. A more representative figure in this respect is the English physician and naturalist Martin Lister (1638?–1712). Almost immediately after his election to the Royal Society in 1671 Lister commented on Steno's newly translated *Forerunner* in a letter sent to Oldenburg for publication in the *Philosophical Transactions*[14]. This has some claim to retrospective fame—or notoriety—as probably the first contribution to palaeontology to be published as a short paper in a scientific periodical.

Lister's rejection of the organic theory of fossils is of particular interest for several reasons. There was no one in Europe, perhaps, in a better position to appreciate the resemblances between the commonest fossils and their living analogues. Lister had already embarked on the major study of molluscs that was to culminate in his great illustrated catalogue of *The History of Shells* (1685–92). Yet both in that work and

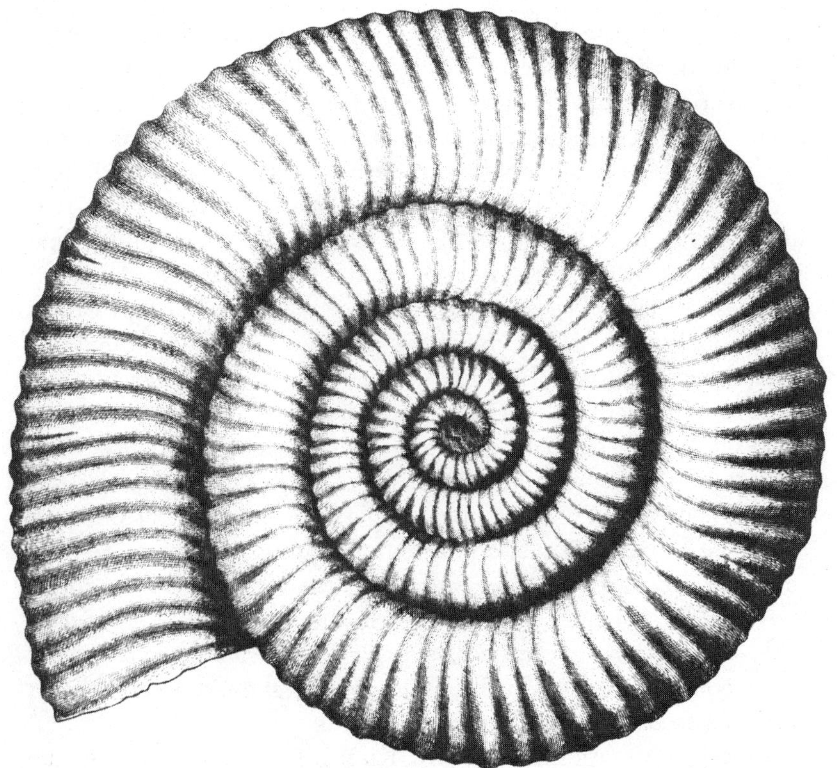

Fig. 2.5. Lister's illustration (much reduced) of a giant ammonite, two feet across (1692)[15]. *Such fossils were so unlike any living molluscs, and so confusingly preserved, that Lister doubted whether they were organic. Hooke, on the other hand, accepted their organic origin but attributed them to a remote age in which most animals—and Man himself—had been of giant size.*

in his earlier *History of English Animals* (1678) he described and illustrated fossil shells in separate sections, terming them shell-stones (*cochlites*, *conchites*, etc.) and denying their organic origin, despite their similarities to the shells of living molluscs[15]. His reasons for this conclusion, stated briefly even in his letter of 1671, illustrate the difficulties that could still face a highly competent naturalist when tackling the problem of fossils.

Lister was prepared to accept Steno's interpretation of his Italian fossils, for these were at the 'easy' end of the spectrum. The English fossils, however, with which he himself was most familiar—in modern terms mostly from Jurassic and Carboniferous strata—raised much greater problems of interpretation. First there was the problem of matter: "there is no such thing as *shell* in these resemblances of shells", he asserted, and indeed many of his fossils were specimens with confusing modes of preservation. Closely connected with this was the problem of their position: "Quarries of different stone yield us quite different sorts or species of shells". The fact that particular fossils are often characteristic of particular strata was to prove at a later period to be the key to historical palaeontology, but to Lister it was a stumbling-block. If fossil shells were truly organic in origin and had merely been 'thrown up' on land as Hooke had suggested, the same kinds should have been found everywhere. Since on the contrary they seemed to be peculiar to particular rock-types and localities, it naturally suggested that they had grown there *in situ*, just as plants grew in characteristic habitats at the Earth's surface. However, Lister's third and most important objection arose from the problem of form: unlike Steno's fossils, "our English Quarry-shells", he said, "were not cast in any *Animal mold*, whose species or race is yet to be found in being at this day" (Fig. 2.5). He knew from his careful systematic work on molluscs that his fossil shells had only a general resemblance to living species, and not an exact identity. Although he did not mention Hooke by name, he criticised those who were content to see general similarities, and maintained that they would be obliged to change their minds "when they shall please to condescend to heedful and accurate descriptions". Lister was even prepared to accept this as the crucial issue on which the question should be decided: if living molluscs were to be found specifically identical with his fossil shells, he concluded, "my argument will fall, and I shall be happily convinced of an Errour". Until then, he maintained, "I am apt to think, there is no such matter, as Petrifying of Shells in the business . . . but that these Cockle-like shells ever were, as they are at present, *lapides sui generis*, and never any part of an Animal".

V

The force of Lister's objections was stated more explicitly by his friend John Ray (1627–1705), perhaps the greatest naturalist of the age[16]. Ray had been on an extensive tour of the Continent a few years earlier —he too had met Steno at Montpellier—and in the course of primarily botanical studies he had noted the principal collections of 'fossils'.

When he came to publish a narrative of *Observations* (1673) on his journey, he used his notes on a German collection as a pretext for an essay on the problem of fossils[17]. He set down, with characteristic fairness, the arguments for and against both the organic and the inorganic theories, quoting Hooke and Lister respectively to illustrate these positions. As a distinguished naturalist he was well prepared to appreciate the arguments for the organic theory—it was, he said, "to me [the] most probable opinion"—yet at the same time he recognised two serious objections to it. Firstly the differences between living and fossil species would imply that some species had become extinct; and secondly the occurrence of fossils on high hills, even on the Alps, was difficult to explain either by invoking the Flood or by attributing the upheaval of mountains to the action of earthquakes. These recurring problems, of form and position respectively, were now becoming increasingly acute, as the debate broadened from the 'easy' Italian fossils to the more 'difficult' specimens found in England and elsewhere.

On the question of *form*, the essential dilemma was that the philosophical presuppositions inherent in the natural theology of the period led to conflicting conclusions. On the one hand, they encouraged an Aristotelian functionalist approach to biology—well exemplified in Ray's own work—which made it literally incredible that the similarities between fossil and living shells should be merely fortuitous. If fossil bivalve shells, for example, were hinged like living bivalves, it seemed absurd to suggest that that structure had not served the same function in both cases. As Hooke and Ray recognised, this teleological view was a powerful argument in favour of the organic theory. On the other hand the frequent lack of exact identity between fossils and living species conflicted with this conclusion, because it implied the probability of extinction. The difficulty was that extinction unavoidably suggested some imperfection and incompleteness in the design of the original Creation. It thereby threatened not only the Christian doctrine of providence, but still more the older notion of plenitude with which that doctrine had become entangled. It seemed inconceivable that any forms of animate being that *could* exist, and that evidently *had* existed, should subsequently have been allowed to disappear from the face of the Earth. This was the objection that lay implicitly behind Lister's opposition to the organic theory of Steno; and it was the same objection that Ray, in his judicious evaluation of the problem, felt as a difficulty in the position he was otherwise inclined to accept.

There was only one way out of this dilemma: the fossil species might not really be extinct at all. Ray felt that it might be more reasonable to assume that they still existed alive somewhere in the world, although as yet undiscovered. Since the organisms under dispute were mostly marine animals, and exotic marine faunas were still poorly known, this was a perfectly justifiable conclusion; indeed for the particular case that Ray cited, the fossil stalked crinoids, it was vindicated half a century after his death by the discovery of living specimens in very deep water in the West Indies.

Yet even Ray himself seems to have felt that this was not a wholly satisfactorily solution of the problem. In a later essay on fossils he singled out the ammonites as those that puzzled him most of all: for here was a whole genus, containing many different species, totally unknown in the living state, and yet Hooke had produced good reasons for thinking that they were organic in origin. So powerfully, however, did Ray feel the force of the objections to extinction, that he was inclined to place the ammonites on the inorganic side of the divide, although he was prepared to accept the organic origin of many other less puzzling fossils. However it is only fair to add that the exceptionally confusing modes of preservation of ammonites provided strong support for his position on this point.

Hooke, on the other hand, was so convinced on teleological grounds that ammonites must have been true shells similar to those of the living nautilus, that he was prepared to accept the corollary that they had become extinct. Yet he was working within the presuppositions of natural theology no less than Ray and Lister, and so it is significant that he was *also* prepared to accept the possibility that new species had arisen in the course of time by some process analogous to the formation of new varieties by breeding. This was formerly considered sufficient ground for hailing Hooke as an enlightened forerunner of Darwin; but in fact it was less an evolutionary theory than a device for maintaining the plenitude of the created order. So long as new species had been formed while others became extinct, it could still be said that the fullness of creation had been preserved, even if the particular species in existence had not always been exactly the same. However, this was not a device that seemed persuasive to many of Hooke's contemporaries, and it was not developed further until a much later period.

The second of Ray's reservations about the organic theory, namely the problem arising from the *position* in which fossils were found, was to occasion active discussion and controversy for the rest of the century. Ray, however, was only expressing an uneasiness that was felt by everyone who considered the origin of fossils at all. No argument for their organic origin, based on their similarity of form to living organisms, could be persuasive unless it was coupled with an explanation of how such organic objects had come to be high above sea-level and embedded *within* rocks.

Scilla's work shows an easy—indeed too easy—solution of this problem. In his frontispiece 'Sense' was shown seated on a hillock strewn with the fossil remains of marine animals, but these fossils were depicted lying loose on the surface, not embedded in strata (Fig. 2.3). Although Scilla was well aware that they did in fact occur within strata, he regarded these merely as current-swept accumulations of sand and gravel, and argued that the emplacement of the fossils had been due to an exceptional series of tidal disturbances, probably at the time of the Flood.

Such a simple solution was not open to Steno, because he was too well aware of the nature of stratification. In his *Forerunner* he therefore elaborated his earlier arguments for the origin of strata by precipitation from a fluid, and showed how the shells of molluscs and the teeth of sharks could have become enclosed within the sediment as it accumulated layer by layer. Now he emphasised that these strata must have been deposited originally in *horizontal* layers. This implied that the tilted position in which they were commonly observed must have been due to subsequent changes. The phenomena could therefore be used to reconstruct, by rational inference, a sequence of events in the Earth's history.

Steno expounded this method of reconstruction by analysing the rocks of Tuscany, but he believed that this area would be found typical of the Earth as a whole (Fig. 2.6). He distinguished two separate periods of horizontal sedimentation in a fluid, two periods in which the underlying strata had been excavated by some subterranean agency, and two periods in which the remaining strata had therefore collapsed. The six periods thus distinguished are important not merely as the first major attempt to reconstruct a sequence of geological events, but even more because the two cycles differed fundamentally from each other and provided the basis for a directional history of the Earth. Steno's earlier strata—the slaty rocks of the Apennines—contained no fossils, and therefore, he believed, antedated the appearance of life on the

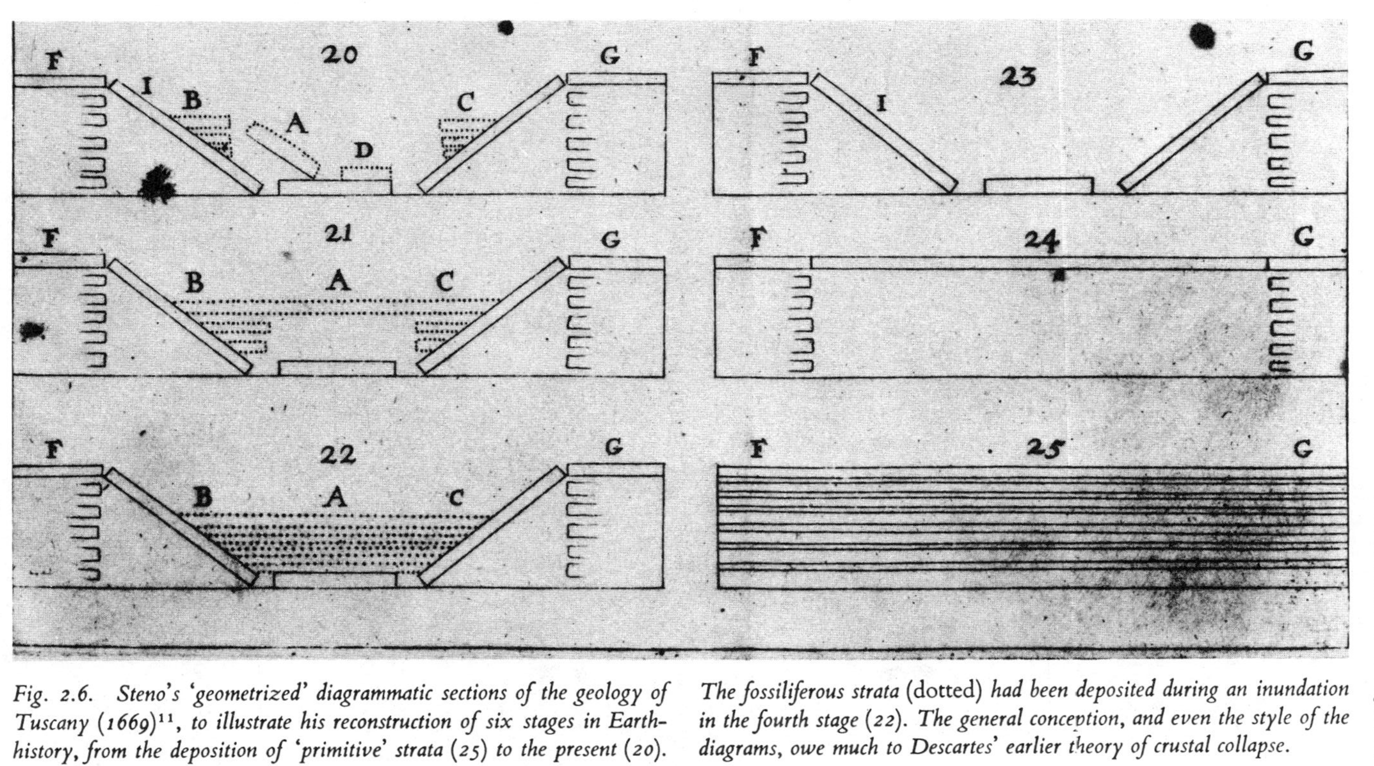

Fig. 2.6. *Steno's 'geometrized' diagrammatic sections of the geology of Tuscany (1669)*[11]*, to illustrate his reconstruction of six stages in Earth-history, from the deposition of 'primitive' strata (25) to the present (20). The fossiliferous strata* (dotted) *had been deposited during an inundation in the fourth stage (22). The general conception, and even the style of the diagrams, owe much to Descartes' earlier theory of crustal collapse.*

Earth; whereas his later strata—the Cainozoic Subappenine rocks—contained what he had shown to be genuine organic remains, and therefore witnessed to an epoch after the creation of life. Here, for the first time in a published work, fossils were used as evidence for the construction of a history of life. This was the new dimension that Steno's studies, prompted initially by the unexpected shark's head, introduced into the debate on fossils.

VI

It is at this point, however, that earlier commentators on the history of geology often felt constrained to turn regretfully from praising Steno as the forerunner of modern geology to apologising for his accommodation of Earth-history within the confines of a few thousand years. Yet it is important to recognise that, in the context of his time, Steno's attempt to harmonise his observations with scriptural history was no insincere or forced reconciliation, but a natural synthesis of what he and his contemporaries regarded as two equally valid and complementary sources of evidence—the Book of God's Word and the Book of God's Works. Thus he could say, of one episode that "Scripture and Nature agree"; of another, that "Nature says not, [but] Scripture relates"; of another, that "neither Scripture nor Nature declares"; and of yet another, that "Nature proves [it], and Scripture does not gainsay it". He had no need to twist his evidence to agree with scripture: on the contrary he sincerely believed that his observations of Nature merely amplified and made more intelligible the bare outlines of prehistory recorded in the early chapters of *Genesis*.

In assessing the seventeenth-century belief that the Earth was only a few thousand years old, historians of science have perhaps been less ready to grant to this limited view of time the historical sympathy and understanding that is now accorded to the limited space of Ptolemaic—and even Copernican—cosmology. Yet the cases are closely parallel: a universe of limited time, like a universe of limited space, seemed at the periods concerned to be supported by almost all the evidence available and by almost all the arguments of good common sense, as well as by those of tradition. It was not simply the force of mental conservatism, still less the threat of ecclesiastical sanctions, that made almost all thinking men agree that the Earth was of quite recent origin—though indeed very ancient by the standards of human life. Even Thomas Hobbes, for example, who was certainly heterodox in many of his opinions and perhaps even an atheist, could take this particular belief for granted as easily as any more orthodox scholar[18]. The problem of fossils was to become one of the points at which this conventional belief would come under strain; but it is only natural that that problem was at first accommodated within what seemed to be a well established conclusion.

Far from being a strait-jacket that cramped the development of an understanding of the significance of fossils, the biblical concept of time had initially the opposite effect[19]. The cosmological debates of the sixteenth and earlier seventeenth centuries had been conducted largely within philosophical frameworks that gave little encouragement to questions of the origin and development of the Earth[20]. Terrestrial changes, although continuous and, on the Aristotelian scheme, taking place on an infinite time-scale, could not be said to be directional any more than the incessant regular movements of the heavens. The idea that the Earth had had a *history*—using that word in its modern sense and not in the earlier sense that is preserved in the term 'natural history'—first entered 'scientific' debate not from the realm of natural philosophy but from that of evidential theology.

A strong sense of linear or directional history had of course been characteristic of Jewish and early Christian thought: indeed it was central to the belief that God was continuously active within history. In its Christian form this belief provided a framework for history extending from the Creation and the Flood, through the Old Covenant with Israel to its fulfilment at the mid-point of history in the coming of Christ. From there it extended through the period of the New Covenant with the Church to the future Judgment and Millenium, and so to the final Consummation of all things[21]. In the sixteenth century this directional concept of history was renewed not only by the conscious attempt to recapture the authentic doctrine of the New Testament, but also by the growing sense that human history was indeed directional even in secular terms[22]. Furthermore, since the secular was not in fact distinguished as a separate sphere, this nascent concept of human progress was commonly linked with expectations of a more or less imminent Millenium, which would see the establishment of a political, social—and scientific—Golden Age transcending even that of Antiquity[23].

With this heightened awareness of being involved in a dramatic historical process, it was natural that there should have been a renewed concern with historical scholarship. Using increasingly sophisticated methods, this led to the critical comparison and correlation of all available historical records, and to the attempted compilation of a unified chronology of world history. The study of Chronology became in the seventeenth century a scholarly pursuit of the highest respectability; and it was one which Isaac Newton, for example, felt worthy of as much serious and detailed research as his work on natural philosophy[24]. Newton also illustrates the importance of apocalyptic prophecy as a motive for the study of chronology: it was by no means a purely antiquarian pursuit, but was often felt to have immediate political and social relevance.

James Ussher, Archbishop of Armagh, who is commonly though erroneously credited with the invention of the date 4004 B.C. for the creation of the world, was only one of many scholarly chronologists who applied the most up-to-date knowledge of comparative philology to the solution of these problems. When Ussher published his *Annals of the old covenant from the first origin of the world* (1650), it was regarded by natural philosophers not as a deplorable specimen of clerical obscurantism but as a product of the finest historical scholarship of the age. Ussher pointed out that whereas ancient authors had all despaired of carrying chronology back to the origin of the world, modern knowledge of astronomical calculation, of the Hebrew calendar and of extra-biblical records now made this problem capable of solution[25]. The date with which his annals began, with the origin of the world on the night preceding Sunday the 23rd of October in the year 710 of the Julian calendar (and 4004 before the conventional date of Christ's birth), was a date with which other chronologists could—and did—disagree in detail; but few of them doubted the validity of his methods of historical analysis or questioned the possibility of arriving at a chronology of some such precision.

Underlying that precision there was, of course, the assumption that a literal interpretation was the only valid exegesis for even the earliest chapters of *Genesis*, and moreover, that those chapters were divinely authenticated as historical records. But this assumption was encouraged by an increasing demand for exegesis that could be seen to be rational, and by a corresponding distaste for allegorical or indeed any non-literalistic methods of interpretation. Kircher's *Noah's Ark* (1675)[26], with its astonishingly detailed analysis of the recorded dimensions and faunal contents of the Ark, is a characteristic specimen of the literalistic and yet also highly erudite work that was produced within this tradition. This renewed literalism seemed to be vindicated, however, by the comparative studies of the chronologists themselves. The Bible was unique not simply as the authoritative source of Christian doctrine: it was also felt to be virtually the only reliable source of information on the earliest periods of history. Even Greek sources, and still more such oriental records as were known, soon lapsed into manifestly legendary and mythical accounts when traced back in time; whereas the ancient Hebrew chroniclers, with their acute sense of the historical, seemed by contrast to be offering sober factual accounts of the history of their own people, and indeed of the human race, back to the beginning of the world.

No doubt this conclusion was reinforced by the feeling that to accept the literal reliability of some of the oriental chronicles would be to set foot on a dangerous slope towards Aristotelian eternalism. Isaac de la Peyrère's notorious work on *Pre-Adamic Men* (1655) would have seemed an apt warning of this danger, all the more because it was intended to preserve the truth of the biblical account and yet had done so only at the cost of denying its universal reference[27]. This fear was clearly an important motive in, for example, Sir Matthew Hale's work on *The Primitive Origination of Mankind* (1677); but at the same time Hale was able to produce many cogent and rational arguments to support his conclusion that human civilisation was not of great antiquity[28]. Since a very long human prehistory was neither suspected nor acceptable, this seemed tantamount to demonstrating the recent origin of mankind. Furthermore, since an earth without men to inhabit it would have seemed, if not purposeless, certainly incomplete, the same evidence could be taken as indirect confirmation of the brief history of the Earth itself.

Once again, in this last assumption there is reflected not only the Christian doctrine of man as the crown of terrestrial creation, but also the concept of plenitude. Whether the Earth had existed from eternity or only for a few thousand years, it was assumed by all that the history of mankind must be broadly co-extensive with the history of terrestrial nature. It was only at a later period that these two kinds of history began to be considered as separate problems. For the seventeenth century they were essentially parts of the same problem, since they were both concerned with the same periods of Antiquity. It is therefore no accident that antiquarian articles are abundantly represented in the early volumes of the *Philosophical Transactions*, or that many of those who were concerned with fossils as possible clues to the history of nature were also actively involved in historical and antiquarian scholarship[29].

VII

The general agreement among seventeenth-century chronologists, that the world was only about six thousand years in age, did not seriously cramp the style of those, like Steno, who were interested in the natural events that had taken place on Earth. Whatever the precise starting-point of world history, chronology provided a framework within which could be fitted not only the events of civil history but also whatever natural events could be inferred. Steno felt no sense of strain in accommodating his six-phase geological history of Tuscany within the bounds of conventional chronology. For example, it seemed natural to him, as to his contemporaries, to interpret a recently discovered skeleton of a (Pleistocene) elephant as the last survivor of those in Hannibal's army; for there were good historical records for that invasion and none for the existence of elephants in Italy at any other period. So far was he from feeling he had to compress his history into an embarrassingly short time-scale, that his arguments show precisely the opposite concern, namely that he might be ridiculed for placing his events at such a great antiquity. To his contemporaries, it seemed marvellous enough that the coins, urns and statues of classical Antiquity had survived their interment for as long as two millenia, and Steno felt he had to meet the objection that more perishable objects such as sea-shells could not possibly have survived even longer periods buried in the earth. This explains his evident pleasure at finding a chain of evidence that irrefutably pushed the antiquity of some of his fossils back almost to the date computed by chronologists for the Flood. He pointed out that the Etruscan city of Volterra, which had been at the height of its power when Rome was founded, contained among its oldest walls blocks of stone incorporating fossil shells; and the city itself was built on a hill composed of strata that contained other shells, not even lithified. This proved, he maintained, that shells could survive almost unchanged for at least three thousand years, so that it was reasonable to attribute the epoch of deposition of fossiliferous strata to the time of the Flood itself. No other event in recorded history had, he believed, been drastic enough in its effects to account for the position of these marine strata high above present sea-level; and nothing about these strata could be held to be contrary to the brief outline account of that event given in scripture.

Yet Steno's assignment of the deposition of all the fossiliferous strata to the time of the Flood did raise problems that were to remain acute throughout the rest of the century. His solution was felt to be unsatisfactory because it was difficult to reconcile with the increasing demands *both* for a literal interpretation of scripture *and* for natural explanations that would be conformable to reason. It was difficult to conceive how an extensive series of strata, containing marine fossils, could have been laid down either within the recorded duration of the Flood or indeed by the *kind* of Flood described in scripture. Furthermore, even if a more elastic exegesis were adopted, the question of the origin and disposal of the waters of the Flood was still left unanswered. The earlier solution, used for example by Calvin and revived by la Peyrère, was that in view of the divine purpose of the Flood its effects could have been limited to the area occupied at the time by man; but this was now felt to be unacceptably in conflict with scripture. If the Flood had been universal, on the other hand, a prodigious volume of water would have been required to cover the highest mountain peaks; yet to postulate the direct creation and subsequent annihilation of this water seemed an arbitrary interruption of God's normal use of secondary causes. It was in this sense that Ray felt that the Flood was an inadequate explanation of how fossils came to be placed within strata high above the sea.

Hooke too was of the same opinion. In 1668—before the publication of Steno's *Forerunner*—he had enlarged the brief comments on fossils already published in *Micrographia* into a long Discourse to the Royal Society[30]. In this he rejected the use of the Flood as an explanation of the position of fossils on land. But this does not reflect any desire on Hooke's part to reject the conventional chronology of Earth-history, still less to attack the attempted concordance of Nature and Scripture. On the contrary, however 'modern' his conclusions may seem in certain respects, they are set, with no more strain than Steno's, within the framework of a few millenia.

Hooke first expanded his earlier arguments for the organic origin of fossils, again using prominently the argument from natural theology that if fossils were *not* organic they would be without purpose. He then elaborated an analogy between fossil shells, as marks of the former extent of the sea, and Roman coins, as marks of the extent of an ancient empire. This was not, as it became in a later period, a purely methodological analogy between one historical science and another: to Hooke it was a matter of building up a chronology of one and the same period of history, by supplementing one source of evidence with another. The task of the "Natural Antiquary" was to amplify, with the aid of fossils, the records used by the student of 'artificial' antiquities. Furthermore, fossils actually had certain advantages relative to more conventional records. They evidently survived long periods of time even better than the pyramids and obelisks of Egypt, the oldest known civilisation; and they were "more legible" than that civilisation's still undeciphered hieroglyphs. Although the supposed antiquity of the Hermetic writings known to the Renaissance had been exploded earlier in the seventeenth century, scholars such as Kircher continued to believe that hieroglyphs embodied the wisdom of a learned age of great antiquity[31]. Such a belief, however, was by no means an old-fashioned curiosity: Newton, for example, was perfectly serious in believing that he was merely re-discovering principles of natural philosophy that had been known to such an age and subsequently forgotten[32]. Hooke too subscribed to the same belief, and valued 'natural antiquities' such as fossils chiefly for the light they might throw on this obscurely known early period of the Earth's history. He did not doubt the reality of the Flood, for it was recorded in many extra-biblical records as well as in *Genesis*, but he judged it an inadequate solution of the problem of fossils. Unlike Steno, Hooke—more literalistically—judged the Flood to have been too brief an event to account for the emplacement of marine fossils in rocks far from the sea. Instead, citing a wealth of historical records, he argued that earthquakes could have accomplished the observed effects.

Hooke's employment of earthquakes as an agency of terrestrial changes, though not original, enabled him to integrate his belief in the organic nature of fossils with his equally firm belief in the brief time-span of the Earth's existence. On the assumption that the Earth had been hot and fluid at its origin, he could argue that earthquakes had been more potent at first, and therefore able to effect more drastic changes in geography than those that had been recorded in more recent times. This concept of directional change in Earth-history accorded well with the common belief that the Earth was gradually decaying from a state of pristine perfection[33]; and Hooke even borrowed from scripture the organismic analogy of the Earth "waxing old". This in turn agreed with the traditional belief, prompted not only by the legends of Antiquity but also by the discovery of large fossil bones, that the Earth in its 'infancy' had supported a race of giants; and Hooke suggested that the giant ammonites of the Portland limestone (Fig. 2.5), so much larger than any known living shellfish, dated from the same period. These and other species found only in the fossil state could have been destroyed, he suggested, by the same catastrophic earthquakes that were obscurely reported, for example, in Plato's legend of a drowned Atlantis. Likewise the pristine Learned Age, the wisdom of which was now known only in corrupt, mythological or undecipherable forms, could have been annihilated by the earthquake that caused the Deluge which Noah alone had survived.

Within this synthesis of natural and human history there was only one feature that would have struck Hooke's contemporaries as surprising, and that was his assertion that the extinction of old species and the formation of new ones during history was "not unlikely". But his use of earthquakes, though an attractive new factor in the debate, was also not wholly satisfactory, as Ray clearly recognised. It was far from certain that historically recorded earthquakes had generally had the effects that Hooke ascribed to them, or that they could account for the position of fossils within strata. Moreover, while accepting the reality of the Flood Hooke left its natural mechanism as obscure as ever. A serious attempt to solve this problem came later from a different source.

VIII

Behind the developmental Earth-history sketched by Hooke and Steno lay another influence, barely acknowledged yet clearly powerful. The renewal of chronology within the framework of the Christian concept of directional time had focussed attention on the pattern of both past and future historical events on Earth; but at the same time the revolutions in cosmology had removed the Earth from its fixed centrality in the cosmos and had set it loose within an apparently centreless and infinite universe. Not only did this make acute—long before the Space Age—the problem of the 'plurality of [inhabited] worlds', with all its attendant metaphysical and theological difficulties, but it also implied that whatever history the Earth had had, and would have in the future, might be only a single example of a pattern common to all such bodies. It was in Descartes' *Principles of Philosophy* (1644) that these implications were set out in a form that clearly influenced both Hooke and Steno.

Within his grand all-embracing scheme for a hypothetical universe, to be derived from the most self-evident principles of a natural philosophy that admitted only matter and motion as fundamentals, Descartes included a brief outline of the development of a hypothetical earth. "I have described this Earth", he said, ". . . as if it were only a machine in which there was nothing whatever to be considered but the shapes and movements *of its parts*"[34]. Descartes' scheme traced the natural development of an Earth—*any* Earth, for he disclaimed any intention of offering more than an hypothesis—from its hot star-like origin to its ultimate fate as a cold planet. This was the notion that Hooke, ignoring Descartes' disclaimer, applied to the problem of the origin of *the* Earth, integrating it into his synthesis of Earth-history. Descartes had also sought to show that subsequent to the origin of an Earth-like body, its originally smooth outer surface would have solidified as a crust and later collapsed irregularly into an underlying fluid layer, thereby producing an irregular and variegated surface. Transposed to the known Earth, this natural mechanism could of course be used to explain such features as oceans, continents, mountains, and tilted strata. Steno applied it in just this way: his mechanism of stratal collapse is closely similar to Descartes' diagrammatic representation, and he explicitly agreed with Descartes that his older (unfossiliferous) strata dated from the first origin of the Earth. Thus both Hooke and Steno were attempting, however tentatively, to use their observations of rocks and fossils to construct histories of the Earth that would conform *both* to the evidence of chronology *and* to Cartesian canons of rationality.

In attempting such a synthesis they were following an approach that had begun almost as soon as Descartes' work appeared. Among those who had welcomed his cosmology—or at least some aspects of it—most wholeheartedly was the 'Cambridge Platonist' philosopher

Henry More, who saw in the infinity of worlds a startling but attractive extension of the plenitude of the created universe[35]. More sketched the implications of the Cartesian scheme not only for the infinity of space but also for the infinity of time. However, he did not draw an Aristotelian conclusion that the Earth itself had existed from eternity, since this was clearly contradicted by scripture and was also improbable on Descartes' hypothesis of cosmic development. So, he wrote,

> I will not say our world is infinite,
> But that infinitie of worlds there be,

an infinity, that is, in time as well as space. He concluded that,

> Long ago there earths have been,
> Peopled with men and beasts before this Earth,
> And after this shall others be again,
> And other beasts and other humane birth.[36]

In this way More was able to reconcile the conventional scriptural chronology for the Earth with Descartes' mechanistic cosmology, and to synthesise both into an enlarged conception of a designful universe. This synthesis was achieved, however, at the expense of depriving the scheme of Christian history of its traditional cosmic significance and reducing it to a purely local account of the origin and destiny of one of an infinity of worlds. Thus More suggested, for example, that the much-discussed *novae* represented distant worlds that had flared up in Conflagrations like that traditionally predicted for the Earth. A further implication was that such dramatic physical events as the Earth's past Deluge and future Conflagration were essentially natural to the history of any Earth-like body.

These implications were drawn out more fully by Thomas Burnet (1635?–1715), who like Ray and Newton was profoundly influenced by the Cambridge Platonists. In his *Sacred Theory of the Earth* (1680, 89)[37] Burnet never mentioned fossils, and indeed his work actually aggravated the problem of accounting for their position; yet indirectly it was the most important influence on the course of the debate about fossils in the late seventeenth century. Burnet, who was a man of extremely wide learning, set out to use all available scientific and historical evidence to amplify scripture and so to make the scriptural events rationally intelligible and intellectually respectable. His theory was 'sacred' in the sense that it focussed attention on the major events in the Christian scheme of history; but its total omission of the traditionally central event, the Incarnation, reflects not only his preoccupation with those events for which physical evidence could be expected, but also the charactistically deistic tendency of his theology. The events he did discuss were the Primaeval Earth, Paradise and the Deluge; and in the future, the Burning of the World, the New Heavens and the New Earth, and the Consummation of all things. Counting the present state of the Earth, this gave a symmetrical series of seven major phases of Earth-history. Like More and Hale, he rejected eternalism for the Earth itself, and indeed framed his argument explicitly to refute such suggestions. His scheme was devised almost entirely within the limits of conventional chronology, which he saw little reason to doubt: like Newton he thought the 'days' of Creation probably represented periods of a year or more, but this made little difference to his time-scale. However, following Descartes, the physical events within that time-scale were attributed to natural causes inherent in the construction of the Earth. Furthermore, the symbolic frontispiece of the book shows that like More he saw the Christian drama of history as a purely terrestrial affair: the figure of Christ proclaimed "I am alpha and omega" from a position astride the first and last of the seven phases of Earth-history, which were shown surrounded by outer space filled with cherubim (Fig. 2.7).

As regards the debate on fossils, the most important part of Burnet's work was that dealing with the Deluge. For this, as for his other events, he tried to find physical explanations that would satisfy the text of scripture and other ancient records, properly understood, and at the same time be framed within the Cartesian philosophy of nature, which permitted explanation only in terms of matter and motion. Thus he argued that the Earth before the Deluge had been smoothly spherical, with an axis of rotation that gave it a climate of perpetual spring. This enabled him to combine the ancient tradition of a paradisal world without seas or mountains with the Cartesian mechanism for the formation of celestial bodies. For the Deluge itself he then argued that the only adequate explanation was to invoke (like Steno) the Cartesian mechanism of crustal fracture and collapse, identifying the fluid layer beneath the crust as the 'great deep' of scripture. All the present topography of the globe was thus the result of the Deluge: man now inhabited a mere "Ruin" or "Broken Globe". This explanation had the

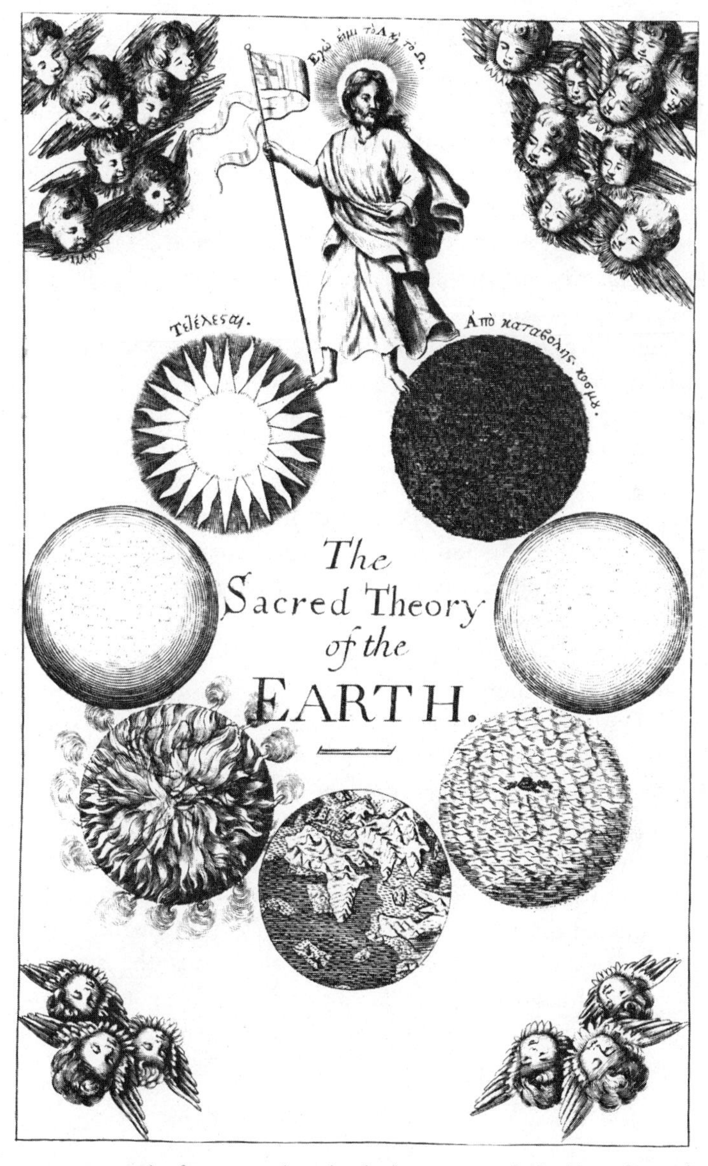

Fig. 2.7. The frontispiece (1684) of Thomas Burnet's highly influential book on Earth-history[37]*. Christ ('I am Alpha and Omega') stands astride the first and last phases of a clockwise sequence in which the Flood had been the third phase (note the Ark afloat), the present was the fourth, and the anticipated Conflagration would be the fifth. Burnet's theory, which won Newton's approval, failed to account satisfactorily for the origin of fossils, but stimulated active speculation and heated discussion of the problem.*

further advantage of reconciling his aesthetic revulsion at the disorderliness of the Earth's topography, and especially mountains—a common enough sentiment at the time—with his equally strong feeling that the Earth must have been intended to be orderly and designful[38].

However, while this was satisfactory for his natural theology, it raised other problems. Of these the exegetical difficulties were least serious: like Newton, who corresponded enthusiastically with him about the theory, Burnet distinguished between 'vulgar' and 'philosophical' senses of scripture, and believed his explanation was confirmed by a correct 'philosophical' interpretation of *Genesis*[39]. More serious was the fact that his explanation of the Deluge could not be reconciled with an organic interpretation of the marine fossils found in strata in mountains, for on his theory these mountains were nothing but the broken fragments of a crustal layer that had formed in a world without seas. Furthermore, as Newton pointed out, the original creation of marine life would have been nugatory, and another episode of creation would have been needed after the Deluge in order to populate the newly-formed oceans. Most serious, however, was the problem arising from the natural inevitability of the Deluge on the mechanism Burnet proposed. If the Deluge had been 'built in' to the Earth's development, it was difficult to see how it could have had a providential function, for the Fall had not yet occurred when the Earth was first designed. In this way Burnet's theory posed in an acute form the question of the nature of God's activity in the created world. Naturalistic explanations of the great physical events of Christian history made their providential status problematical; yet supranaturalistic explanations were postulated only with reluctance, because they infringed the basic principles of *both* Puritan covenant theology *and* the newer deistic theology, by throwing doubt on both the reliable constancy and the rational intelligibility of God's relation to the world.

IX

With a fine sense of timing, Burnet published the second half of his *Theory*, dealing with future events, shortly after the Glorious Revolution, to which many of his contemporaries gave an apocalyptic interpretation, and on the eve of a decade that had long been predicted as one of special apocalyptic significance. It is symptomatic of the reasons for interest in historical questions that it was this part of Burnet's work that provoked lively controversy, whereas the first part had created little stir in spite of its speculative implications. The crucial issue was whether the future Millenium, or whatever physical event might presage it, were events 'built in' to world history and therefore in principle capable of being predicted; or whether their providential status was to be preserved by denying their inevitability.

It was in this context of apocalyptic speculation that the debate about fossils was revived again in England in the 1690s. Ray's first reaction to the growing discussion of Burnet's theories was to publish a sermon on the reality of the apocalyptic future, which he had delivered at Cambridge many years before, during an earlier period of apocalyptic excitement. But the section dealing with the possible built-in mechanisms for a future 'Dissolution' was now enormously expanded. In considering the possibility that the Deluge might be repeated, he was led to discuss its first occurrence; and this in turn gave him an excuse for including a long essay on fossils, as possible evidence for the first Deluge. Later, finding these (very) *Miscellaneous Discourses* (1692) well received, Ray re-cast the material in a more orderly form as *Three physico-theological Discourses* (1693) on the Creation, the Deluge and the Conflagration, indicating by this arrangement his intention to provide a more satisfactory alternative to Burnet's work[40].

As in his earlier essay on the subject, Ray set out with conspicuous fairness the conflicting arguments on the nature of fossils. On that occasion he had mentioned the possibility that some fossils might be organic and others not; "yet methinks", he had said, "this is but a shift and a refuge to avoid trouble, there not being sufficient ground to found such a distinction". Now, however, he found himself forced towards such a compromise solution. For some fossils, such as the Maltese tongue-stones, an organic origin seemed inescapable; but for others, such as ammonites and the flattened plants on the surfaces of some pieces of coal-shale (Fig. 2.8), he inclined to believe that "Nature doth sometimes *ludere* [that is, play], and delineate Figures" purely for ornament: while he had no hesitation in grouping the inorganic-looking belemnites along with marcasite nodules as objects that were clearly on the inorganic side of the divide.

However, having accepted an organic origin for at least *some* fossils, Ray had still to account for their position. He could not adopt Burnet's view that the antediluvian Earth had been without seas and mountains.

This would have conflicted too sharply with his belief that the present form of the world was still as designful as before the Deluge—he included an essay on the usefulness and designfulness of mountains as an answer to Burnet on this point—as well as with his belief that some fossils were the real remains of marine organisms. On the other hand he could and did borrow from Burnet the view that the Deluge had been due primarily to an overflow from the 'great deep' within the Earth; but he utilised the traditional notion of a subterranean connection between the oceans and inland springs as a device to explain how, during this event, marine organisms could have been violently transported underground from the seas up on to the land.

Ray's tentative explanation of the transport of fossils from the sea on to the land during the Deluge was ingenious but hardly satisfactory. As he himself must have realised, it could not explain the position of fossils *within* strata. This was the deficiency that the physician John Woodward (1665–1728) sought to make good, in his *Essay toward a Natural History of the Earth* (1695)[41]. Woodward placed great emphasis on his own excellent qualifications for the task; and while modesty was not one of his virtues, it is true that he had a wider first-hand knowledge of rocks and fossils than any of the other writers of the period. In particular he was well aware of the nature of the strata in which fossils were found, and he saw that these were essential to any satisfactory solution of the problem. Without acknowledging his debt to Steno—though one of his critics made it explicit[42]—Woodward framed his theory around the postulate that all fossiliferous strata had been laid down horizontally at the time of the Deluge. The fossils they contained dated from the ante-diluvian period. Together with all the materials of the Earth's surface, they had been churned up into a kind of suspension at the time of the Deluge, "their constituent Corpuscles all disjoyned, their Cohaesion perfectly ceasing". From this thick suspension these materials, and the fossils, had then settled out in order of their specific gravity, to form the observed order of strata with their characteristic fossils. The strata had subsequently collapsed into tilted positions, but in general the post-diluvial world was one of ordered tranquillity.

This theory enabled Woodward to combine an organic interpretation of as many fossils as he wished, with a belief in a Deluge as universal and as catastrophic as Burnet's. The problem of form in fossils did not worry him, because he felt confident in adopting Ray's tentative suggestion that extinction might be only apparent: since little was known about deep-water faunas, it was "very reasonable", he suggested, to conclude "that there is not any one entire species of Shell-fish, formerly in being, now perish'd and lost". While dramatising the Deluge in Burnet-like terms as "the most horrible and portentous catastrophe that Nature ever saw", he also emphasised the tranquil orderly designfulness of the world both before and after that event. Even the Deluge itself was brought within the same beneficent teleological view, not now as a punitive event but as a means of reforming the Earth into a physical state more suited to fallen men. However Woodward did not feel that this was taking unacceptable liberties with scripture: on the contrary he gave the concordance of nature and scripture as one of his chief motives for writing the work. He clearly envisaged all the events he described as having taken place within the bounds of conventional chronology: he suggested, for example, that the Ancients had found fossils more abundantly than in his own day, because they had lived closer to the time of the Deluge and the fossils had had less time to decay away. Unlike Hooke, on the other hand, he rejected any idea of an ante-diluvian learned Age, and he dismissed hieroglyphs as records of a post-diluvial age too primitive to be of any historical value.

X

Woodward criticised his predecessors for the "shortness of their Observations"; but, as Ray commented with unusual asperity, this was just the test which Woodward's own theory was unable to meet[43]. It was clear to anyone who studied the order of strata and their contained fossils that they were *not* arranged in order of specific gravity. For all its use of the latest scientific jargon of corpuscles and gravitation, Woodward's theory failed the simplest empirical test. Moreover, it was highly unsatisfactory in that it failed to suggest any mechanism for the Deluge. Woodward denied that he was in any way obliged to produce such a mechanism, arguing that it was sufficient to prove that the event *had* occurred; but while this position was justifiable methodologically, his main reason for adopting it was that he doubted whether any natural cause could be found. In his concern to rebut the atheistic implications of Burnet's use of "an accidental Concourse of Natural Causes" for the Deluge, he found himself with no alternative but to postulate the "Assistance of a *Supernatural Power*" to account for it. Like the future Conflagration, the Deluge thus became an inexplicable interruption in an otherwise orderly natural world.

Thus although Woodward provided an all-embracing explanation that favoured the organic origin of fossils, the acceptability of this theory was greatly diminished by its failure to satisfy the canons of rational explanation—and by the arrogant tone in which it was proposed. Yet other attempts to explain a Burnet-type Deluge in scientific terms fared little better: William Whiston's *New Theory of the Earth* (1696), for example, tried to harness the recent explanation of cometary motions to provide a mechanism for both Deluge and Conflagration; but when subjected to detailed criticism in the light of Newtonian physics it proved almost as inadequate as Burnet's theory[44].

It is hardly surprising, therefore, that Ray found himself more "irresolute" than ever about the whole question of the nature of fossils and their emplacement. This was partly because his friend Edward Lhwyd (1660–1709), Keeper of the Ashmolean Museum in Oxford and a distinguished naturalist, philologist and antiquarian, was developing a theory of fossils that Ray found an attractive "middle way" between the organic and inorganic explanations. Among theories of the generation of organisms, 'animalculist' notions of the embodiment of specific characteristics within the 'seed' of each species had received new support from microscopical studies of spermatozoa and pollen. Lhwyd noticed that most fossils resembled organisms that spread their 'seed' externally—he discounted the few (for example the bones of quadrupeds) that resembled the parts of viviparous animals. He therefore suggested, in a theory that recalls some earlier Aristotelian explanations of fossils, that most fossils had grown *in situ* within the rocks, not as mere imitations of organisms, but from the same 'seed' as the living organisms they resembled, the 'seed' having been washed into those positions through crevices in the rocks[45]. This theory enabled him to avoid *both* the teleological objections to the purely inorganic explanation that Lister had proposed, *and* the grave difficulties of the organic explanation of Woodward.

Ray found Lhwyd's theory attractive, but he was clearly loth to believe that fossils with minutely detailed organic resemblances were not the remains of organisms that had been truly and fully alive at the Earth's surface. The old problem of form, however, remained as acute as ever: indeed it became steadily more acute, as the collecting activities of men like Lhwyd and Woodward enlarged the range of well-

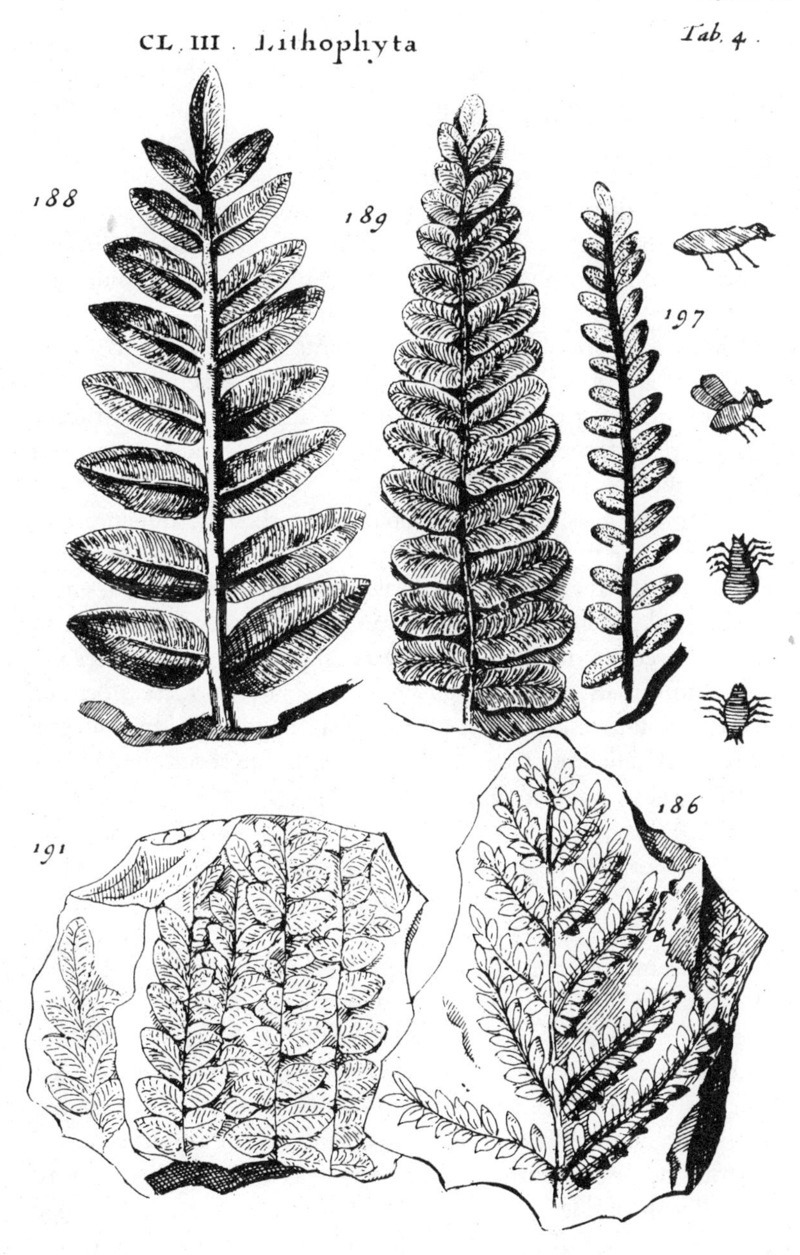

Fig. 2.8. Some of Lhwyd's illustrations (1699) of fossil plants ('ferns' from the Coal Measures)[45]*. He was extremely uncertain about their origin, since no plant* material *seemed to be involved—the fossils were mere impressions—and their* form *differed in detail from that of any known living plants.*

preserved fossil specimens that were clearly different in form from any living species (Fig. 2.8). Furthermore, Ray now felt that such fossils, if truly organic in origin, could not be explained in terms of any catastrophic universal Deluge; and he inclined instead towards suggestions that an Aristotelian theory of the continual slow interchange of land and sea would be more satisfactory[46]. But this placed him in a dilemma: if he accepted the organic theory of fossils, it would throw doubt *both* on conventional chronology *and* on the notion of plenitude. As he wrote to Lhwyd, "On the other side there follows such a train of consequences, as seem to shock the Scripture-History of the novity of the World; at least they overthrow the opinion generally received, & not without good reason, among Divines and Philosophers, that since the first Creation there have been no species of Animals or Vegetables lost, no new ones produced[47]".

This famous passage indicates once more that the difficulties were not simply those of conventional religious orthodoxy; they arose from the combined witness of *both* chronologist scholars *and* natural scientists (if 'philosophers' may be translated anachronistically into that term), a witness moreover that Ray knew was "not without good reason". It was a serious matter intellectually to run counter to such a formidable consensus of the best learned opinion of the age, and it is not surprising that Ray remained uncertain to the end of his life about the solution of the problem.

XI

It might seem, therefore, that the question of the nature of fossils was still as confused after forty years of active discussion in the Royal Society circle in England as it had been before Steno first turned his attention to it. Moreover, this apparently inconclusive debate hardly seems to support the claim that Steno's work broke the earlier stalemate and led to a new dimension in the problem. In fact, however, much of the work published during the following century can be seen as the development of ideas and methods first formulated in the age of Steno, Hooke and Ray; and Steno's work in particular did survive, though scarcely recognised, to become the foundation for a new approach to the history of life. Steno's own publications were virtually forgotten; but his understanding of strata as the records of a *sequence* of events in Earth-history survived in plagiarised form in Woodward's theory to become immensely influential in the early eighteenth century.

As in other branches of natural science, so also on the problem of fossils England became an intellectual backwater soon after the beginning of the century; but on the Continent it was Woodward's theory, more than any other, that provided the grounds for the acceptance of an organic explanation for an increasingly wide range of fossils. For all the flaws of Woodward's system, it did provide a moderately convincing explanation of organic resemblances in fossils, and therefore encouraged the careful description and documentation of fossil remains. To ascribe them to a Deluge-event, however widely that event had to differ from the Flood recorded in *Genesis*, seemed to be moderately consonant with both scripture and reason.

The desire to demonstrate the reality of the Deluge from impeccably rational evidence thus became a powerful motive in the description of fossils and in their organic interpretation. This can be seen, for example, in the work of the Swiss naturalist and physician Johann Scheuchzer (1672–1733), who became an enthusiastic convert to Woodward's viewpoint and translated his *Essay* into Latin to give it wider currency on the Continent. In his *Complaints and Claims of the Fishes* (1708), written probably to counter Karl Lang's use of Lhwyd's theory for the interpretation of Swiss fossils, Scheuchzer made the fossil fish of Oeningen speak up to defend their own organic origin against those who denied they had ever been alive, and to emphasise their witness to reality of the Deluge[48]. Likewise in his *Herbarium of the Deluge* (1709) the same motive can be seen behind his careful descriptions and illustrations of fossil plants (Fig. 2.9)[49]. The continuing problems of discriminating between the organic and the inorganic are indicated, however, by his inclusion of some dendritic markings along with many fine specimens of genuine fossil plants. If his enthusiasm for finding relics of the Deluge misled him here, it led him much further astray when he later described the skeleton of a large Cainozoic amphibian as that of *A Man, a Witness of the Deluge and divine Messenger* (1725)[50]. Prefixing this tract—the term is appropriate—with the motto "Take Heed!", Scheuchzer's didactic intentions here completely outran his judgement and knowledge as a physician, for his own illustration ought to have made it patently clear that his specimen was not man-like at all, whatever else it might be.

Fig. 2.9. An engraving of the Flood, from the title-page of Scheuchzer's book on fossil plants (1709)[49]: note the shells being cast ashore in the foreground. Until the mid-18th century, the Flood remained almost the only reasonable causal agency to which fossils could be attributed, if their organic origin was to be upheld.

Yet if the evidence was sometimes strained to provide support for the reality of the Deluge, it could be equally strained in the opposite direction by those who, in the name of Enlightenment, wished to deny that any such inexplicable event had ever occurred. For example Voltaire, whose first-hand knowledge of fossils was probably minimal, nevertheless felt himself qualified to assert that they gave no evidence of any interruption of the Newtonian regularity of the universe[51]. To reach this conclusion he was obliged to dismiss fossils variously as inorganic productions, as the relics of freshwater lakes, and as shells dropped on land by pilgrims: but these were arguments hardly calculated to persuade naturalists who knew that many fossils closely resembled marine organisms and yet were embedded within strata. On the whole, therefore, it was the diluvialists whose work most encouraged the acceptance of an organic interpretation of fossils.

Such an interpretation was favoured, moreover, by much of the detailed study of fossils that stemmed from the acceptance of Woodward's theory. Belemnites, for example, so long the most 'difficult' of all common fossils to bring within the organic interpretation, yielded at last to a careful comparison between the best preserved specimens and the available living analogues. Lhwyd had noticed that some specimens contained a distinctive structure within a conical hollow at one end of the 'guard', but it was Balthasar Ehrhart who first saw that this structure was a chambered shell analogous to that of the living cephalopod molluscs *Nautilus* and *Spirula*. Combined with his careful analysis of the mode of growth of the 'guard', this structural analogy made the organic origin of belemnites almost indisputable[52]. Such an example was bound to encourage those who believed that many other fossil objects would likewise be found to be the remains of organisms. Furthermore, the vexing possibility of extinction no longer seemed a legitimate reason for denying the organic nature of such fossils, not because extinction had become any more acceptable, but because the exploration of the sea in distant parts of the world gave increasing grounds for doubting whether any fossil species were in fact extinct. The discovery of living stalked crinoids has already been mentioned as vindicating Ray's suspicion that this might be found an adequate explanation of the problem.

General theory and detailed observation thus combined to favour an organic interpretation of an increasingly wide range of fossil objects. In a century that gave great scientific credit to works of description and classification, there was a profusion of local monographs and general treatises on fossils; and in these the gradual assimilation of fossils and living organisms into a single systematic scheme is shown by the way that special nomenclatural suffixes for fossils (for example *-ites* in *conchites*, *ichthyolites*, etc.) fell into disuse. Alternative interpretations of fossils died slowly in the early part of the century, becoming restricted to a narrowing range of objects, until it was recognised that the resemblance between (for example) some concretions and organic structures was a trivial phenomenon that had no bearing on the origin of true organic fossils.

The famous episode of the hoax played on Johann Beringer at Würzburg in the 1720s, far from being typical of eighteenth-century opinions on fossils, is a bizarre manifestation of a debate that was by that time dying. The 'planting' of artificial fossils, moulded to resemble insects, birds, comets and other objects, was no light-hearted student prank but a sordid conspiracy motivated by academic jealousy[53]; and the success of the deception reflects not so much Beringer's credulity as his genuine puzzlement at a kind of fossil that had not been described by any of his predecessors. The discovery of these strange specimens led him to review systematically all previous theories about fossils; and since they seemed (correctly) to be only 'imitations' of organisms he concluded that they added weight to all the earlier arguments for the inorganic origin of fossils. But by 1726, when Beringer published his work[54], such a conclusion was already old-fashioned; and his personal humiliation when the hoax was recognised may well have hastened its final disappearance.

XII

Woodward's adaptation of Steno's ideas, powerfully promoted on the Continent by naturalists such as Scheuchzer, did more than encourage an organic interpretation of fossils. The theory of a diluvial origin of fossiliferous strata, for all its limitations, also served to focus attention on questions of stratification, and hence on the problems of reconstructing the history of the Earth. Woodward's explanation of stratification in terms of specific gravities was easy enough to disprove empirically, but it had been a serious attempt to account for the distinctive order of strata and their characteristic fossils. Even when the explanation was rejected, Woodward's work thus acted as an incentive to study these phenomena more closely. Both Woodward and Steno had emphasised that strata with fossils must have been deposited sequentially at the time of the Deluge; it was a small step to retain this concept of strata but to detach it still further from the notion of the Deluge.

Thus during the eighteenth century there developed a generally agreed view on the classification of rocks, in which the acceptance or rejection of a diluvial interpretation made little *practical* difference. The apparently ancient unfossiliferous rocks, typically outcropping in mountain regions and often unstratified, were termed 'Primary', primitive or ore-bearing; and were attributed to the original consolidation of the Earth's crust, or to the period before the Deluge. The stratified fossiliferous rocks, typically outcropping in lower hills, were termed 'Secondary' or stratified, and were attributed to a later era, or to the Deluge itself. The irregular 'superficial' deposits, typically confined to the lowest ground and generally unconsolidated, were termed 'Tertiary' or alluvial, and were attributed to a relatively recent or post-diluvial period. In practice this tripartite division was little affected by the presence or absence of 'diluvial' labels[55].

This synthesis, which appears in the work of many writers in the eighteenth century, was the lineal descendant of Steno's interpretation not only through the medium of Woodward's theory, but also through that of the great natural philosopher Leibniz. Although a brief summary of his essay *Protogaea* was published even before Woodward's *Essay*, Leibniz's work remained in manuscript until many years after his death. Having been commissioned to write the history of a princely family, he began with what he subtitled "A dissertation on the original form of the Earth and on the vestiges of its most ancient history in the very monuments of nature"[56]. This epitomises neatly the way in which Leibniz, like Steno, whom he greatly admired, wished to integrate Earth-history and human history into a single narrative. Adopting a Cartesian explanation of the origin of the Earth as an incandescent globe, Leibniz postulated the consolidation of an original crust, the condensation of an initially universal ocean, and the subsequent deposition of a sequence of strata containing fossils, with the simultaneous diminution of the ocean by evaporation. The bulk of the essay was in fact devoted to the description and illustration of fossils, and to the demonstration of their organic origin, as a crucial part of his whole synthesis (Fig. 2.10).

The posthumous publication of *Protogaea* in 1749 proved highly influential, for it provided a model of Earth-history that allowed for the organic origin of fossils, preserved Steno's and Woodward's understanding of strata as sequential deposits, and was conformable to both scripture and reason. Its most important effect, however, was to make it possible for the different fossils embedded in successive strata to become evidence of the history of life itself, although that conclusion was not at first drawn in any detail.

Leibniz, like most of his contemporaries, saw no reason to doubt the scale of conventional chronology. Ray on the other hand, as we have seen, recognised that fossils might force the enlargement of that scale. More significantly, Ray foresaw, however dimly, that the question of the antiquity of the Earth might be separated from that of man's origin; that with an appropriate interpretation of the 'days' of Creation the Earth's history might be greatly extended without disturbing the question of man. The difficulty with that solution was that any such extension of the terrestrial time-scale tended to be interpreted as an advocacy of Aristotelian eternalism: as in an earlier period it was not the mere magnitude of time that was felt to be threatening, but the

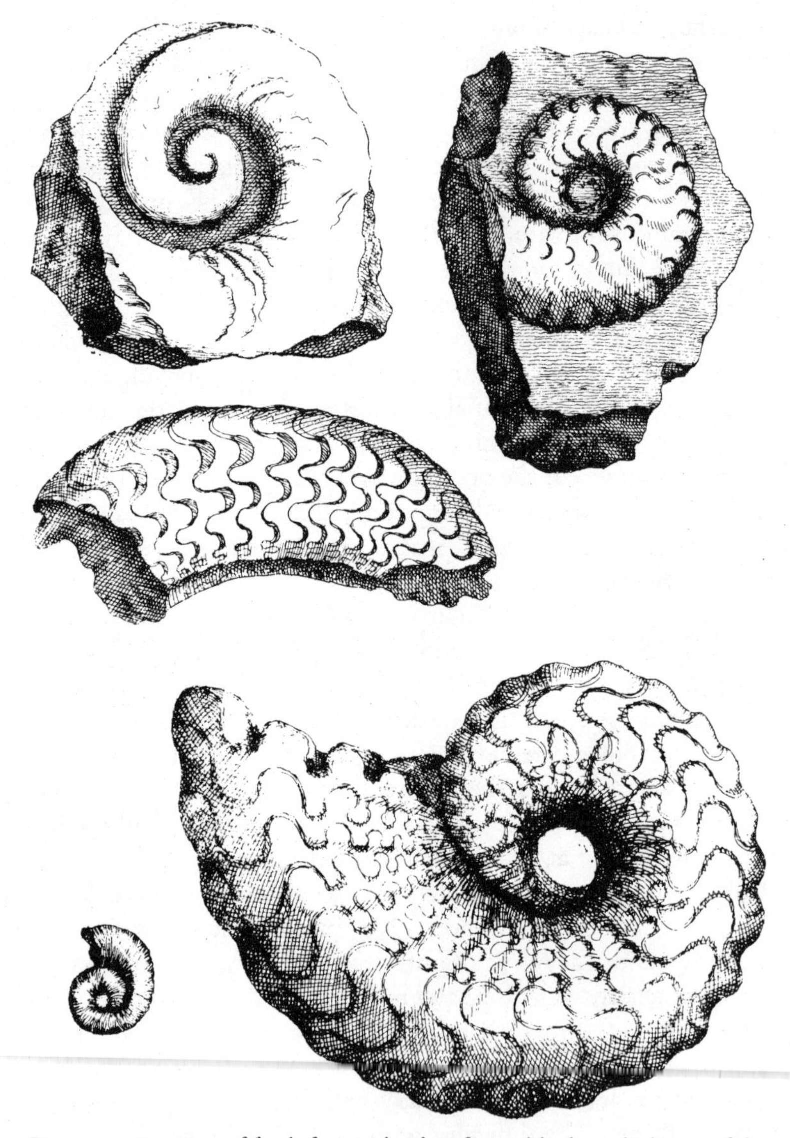

Fig. 2.10. Drawings of fossils from Leibniz's influential book on the history of the Earth (completed in the 1690's, unpublished until 1749)[56]. *The fossils shown here are the distinctive ammonites* Ceratites *from the German Triassic Muschelkalk; in 18th-century terms, from the 'Secondary' formations or Flötz-gebirge.*

concept of eternity. This crucial distinction is nicely illustrated by a paper by the astronomer Edmund Halley, published a few years after Ray's death, in which he proposed an experiment to estimate the minimum age of the Earth from the saltness of the sea, and recognised. that it implied an age far exceeding that of conventional chronology; yet he gave as his reason for the paper the desire to refute suggestions that the Earth could have existed from eternity[57]. It was the taint of eternalism, not the literal interpretation of *Genesis*, that made both 'Divines and Philosophers' hesitant to enlarge the time-scale of Earth-history. Further, there was a rational and imaginative barrier to be overcome. There was still no compelling evidence for a vast time-scale, and anyone who suggested one of millions of years was more likely to be ridiculed than persecuted for his opinion. Thus the clerical editor of Benoît de Maillet's posthumous *Telliamed* (1748) toned down the suggested time-scale of this highly speculative system[58]; but since he left it far in excess of conventional chronology this was more probably to avoid ridicule than to make it appear conformable to scripture.

XIII

It was in the work of the greatest naturalist of the eighteenth century, Georges Buffon (1707–88), that the inheritance of all these earlier writers found its fullest expression. Buffon began his monumental *Natural History* with a volume on the Earth (1749)[59], in which he insisted on the need to find causes of Newtonian regularity for even the most puzzling geological phenomena; and he subjected the systems of Burnet, Whiston and Woodward to scathing criticism for their speculative tone and for their attempt to find concordance between Nature and scripture. Without actually denying the reality of the Deluge, Buffon insisted that it had been moral in its purpose and as miraculous in its effects as in its cause: for scientific purposes, therefore, it could be ignored. Yet although he emphasised that existing physical processes were sufficient to explain all the geographical changes indicated by the position of fossils on land, Buffon set these processes within the limits of conventional chronology, and postulated the original 'softness' of the Earth's materials as an explanation for the large effects that these processes had had within this short period.

However, a quarter-century later, when he came to write a supplementary volume revising this part of his work, Buffon had assimilated Leibniz's work, and he used it as the basis for a reconstruction of the *Epochs of Nature* (1778)[60]. With the inspiration of Leibniz's notion of a cooling Earth, he had conducted experiments with model globes in order to gain a quantitative estimate of the duration of Earth-history. These provided him with concrete physical arguments for a time-scale of tens of thousands of years, though he suspected that only millions of years would suffice for the deposition of all the observed strata. Within this expanded time-scale, Buffon described seven 'epochs'—preserving the form if not the substance of the scriptural account of Creation. The significance of this narrative of the Earth's development lies less in its enlargement of conventional chronology than in its clear detachment of human history from the history of life—or rather in its relegation of human history to a small part, albeit the culmination, of a much longer sequence of events.

Furthermore, within the pre-human epochs, the evidence of fossils was used extensively to reconstruct a true history of life. The 'primitive mountains' were relics of the primary cooling of the Earth's crust, before the first appearance of life. The thick strata of the 'secondary mountains', including many limestones and abundant fossils, bore witness to the early periods of the existence of life; and the giant size of many of the fossils, for example some ammonites, indicated the tropical temperature then prevailing even in temperate latitudes. Later still, with the emergence of the continents from beneath the falling level of the oceans, there had been a period in which elephants and other tropical mammals had flourished in temperate and even subarctic latitudes, leaving their bones to be discovered in the superficial deposits of Northern Europe, North America and even Siberia.

This was an inspiring panorama indeed; but it was precisely as inspiration, rather than as a source of detailed ideas on fossils or anything else, that Buffon's *Epochs* exercised their greatest influence. By the time his work was published, the whole enterprise seemed to a new generation of naturalists to be outdated if not mistaken. There was a growing distaste for grand syntheses of any kind, and a new emphasis on the need to draw only limited conclusions from carefully recorded local observations.

Buffon's system thus lay near the end of an old tradition, not at the start of a new one. It represented the culmination of an endeavour that first impinged on the problem of fossils in the work of Steno and Hooke, namely the attempt to use fossils as one source of evidence for reconstructing and explaining the whole history of the Earth. This had begun, naturally enough, with the attempt to integrate that history into the available records of human history, and there had been no reason to suspect that the two histories might not be co-extensive. It ended with the recognition that historical records were useful only as evidence for the most recent period of Earth-history; that beyond the epoch of man there stretched a long—perhaps unimaginably long—earlier development, which in principle it was possible to reconstruct by deciphering the clues of rocks and fossils. The student of such phenomena was still, as Buffon emphasised, a 'natural antiquary' using many of the same methods of research as the historian; but he was no longer concerned, as Hooke had been, with the same period of Antiquity alone, but with a far longer history of which no man had been witness.

REFERENCES

1. See Gustav Scherz, 'Vom Wege Niels Stensens. Beiträge zu seiner naturwissenschaftlichen Entwicklung', *Acta Historica Scientiarum naturalium Medicinalium*, Kopenhagen, vol. 14, (1956); also 'Nicholaus Steno's life and work', *ibid.*, vol. 15, pp. 9–86 (1958).

2. Olao Wormius, *Museum Wormianum, seu Historia Rerum Rariorum, tam Naturalium quam Artificialium, tam Domesticarum, quam Exoticarum, quae Hafniae Danorum in Aedibus Auctoris servantur*, Lugduni Batavorum, 1655.

3. Nicolaus Stenonis, *Elementorum Myologiae Specimen, seu Musculi descriptio Geometrica. cui accedunt Canis Carchariae dissectum Caput, et Dissectus Piscus ex Canum genere*, Florentiae, 1667. The essay on tongue-stones is incompletely reprinted and translated in Axel Garboe, *The earliest geological treatise (1667) by Nicolaus Steno (Niels Stensen), translated from Canis Carchariae Dissectum Caput*, London, 1958.

4. *Philosophical Transactions of the Royal Society*, vol. 2 (no. 32), pp. 627–8 (10 Feb. 1667/8).

5. V. A. Eyles, 'The influence of Nicholaus Steno on the development of geological science in Britain', *Acta Historica Scientiarum naturalium Medicinalium*, vol. 15, pp. 167–188 (1958).

6. Richard Waller, *The Posthumous Works of Robert Hooke, M.D., S.R.S., Geom. Prof. Gresh. &c. Containing his Cutlerian Lectures, and other Discourses read at the meetings of the illustrious Royal Society*, London, 1705 (facsimile reprint, New York, 1969): see undated discourse, pp. 329–350.

7. R. Hooke, *Micrographia: or some Physiological Descriptions of Minute Bodies made by Magnifying Glasses, with Observations and Inquiries thereupon*, London, 1665 (facsimile reprint, New York, 1961): see 'Observ. XVII', pp. 107–112.

8. See Arthur O. Lovejoy, *The Great Chain of Being. A Study in the History of an Idea*, Cambridge (Mass.), 1936 (reprinted 1960), ch. 5.

9. Athanasius Kircher, *Mundus Subterraneus in XII Libros Digestus; quo Divinum Subterrestris Mundi Opificium . . .*, Amsterodami, 1664–5.

10. Agostino Scilla, *La Vana Speculazione disingannata dal Senso. Lettera risponsiva circa i corpi marini, che petrificati si trovano in varii luoghi terrestri*, Napoli, 1670. Several Latin editions were published in the 18th century.

11. Nicholaus Stenonis, *De Solido intra Solidum naturaliter Contento Dissertationis Prodromus*, Florentiae, 1669. For a modern English translation, see John Garrett Winter, *The Prodromus of Nicolaus Steno's dissertation concerning a solid body enclosed by process of nature within a solid*, New York, 1916 (reprinted 1968).

12. Nicolaus Steno, *The Prodromus to a Dissertation Concerning Solids Naturally Contained within Solids. Laying a Foundation for the Rendering a Rational Accompt both of the Frame and the several Changes of the Masse of the Earth, as also of the various Productions in the same. English'd by H.O.*, London, 1671.

13. Robert Boyle, *Essays of the Strange Subtilty Determinate Nature Great Efficacy of Effluviums . . . Also an Essay, about the Origine and Virtue of Gems. To which is added The Prodromus to a Dissertation concerning Solids naturally contained within Solids Giving an Account of the Earth and its Productions. By Nicholas Steno. English'd by H.O.*, London, 1673.

14. Martin Lister, 'A letter . . . confirming the Observations in No. 74. about Musk scented Insects; adding some Notes upon D. Swannerdam's book of Insects, and on that of M. Steno concerning Petrify'd Shells,' *Philosophical Transactions*, vol. 6 (No. 76), pp. 2281–4 (22 Oct. 1671).

15. Martinus Lister, *Historia Animalium Angliae tres Tractatus. Unus de Araneis. Alter de Cochleis tum terrestribus tum fluviatilibus. Tertius de Cochleis marinis. Quibus adjectus est Quartus de Lapidibus eiusdem insulae ad Cochlearum quandam imaginem figuratis. Memoriae et Rationi*, Londini, 1678. *Historia Conchyliorum*, Londini, 1685–92.

16. Charles E. Raven, *John Ray Naturalist. His Life and Works*, Cambridge, 1942.

17. John Ray, *Observations topographical, moral and physiological, made in a Journey through part of the Low Countries, Germany, Italy and France: with a catalogue of plants not native of England*, London, 1673: see pp. 113–131.

18. Raven, *John Ray*, p. 454.

19. See Francis C. Haber, *The Age of the World: Moses to Darwin*, Baltimore, 1959.

20. Suzanne Kelly, 'Theories of the Earth in Renaissance Cosmologies', *in* Schneer, *Toward a History of Geology*, pp. 214–225.

21. See Oscar Cullmann, *Christus und die Zeit*, Zurich, 1946 (English translation, *Christ and Time. The Primitive Christian Conception of Time and History*, London, 1951).

22. See R. G. Collingwood, *The Idea of History*, Oxford, 1946.

23. See Ernest Lee Tuveson, *Millenium and Utopia. A Study of the Background of the Idea of Progress*, Berkeley and Los Angeles, 1949 (reprinted New York, 1964).

24. See Frank E. Manuel, *Isaac Newton, Historian*, Cambridge, 1963.

25. Jacobus Usserius, *Annales Veteris Testamenti, a prima mundi origine deducti: una cum rerum asiaticarum et aegypticarum chronico, a temporis historicis principio usque ad Maccabaicorum initia producto*, Londini, 1650: see *Praefatio*. I have translated *testamentum* as 'covenant', to emphasise that the work was concerned with a *period* of universal history, not with the Old Testament as such.

26. Athanasius Kircherus, *Arca Noe in tres libros digesta, sive de rebus ante diluvium, de diluvio, et de rebus post diluvium a Noemo gestis*, Amsterodami, 1675.

27. [Isaac de la Peyrère], *Men before Adam, or, a discourse upon Romans V, 12, 13, 14, by which are prov'd, that the first men were created before Adam*, London, 1656 (original Latin edition, *Praeadamitae*, 1655). See also Allen, *Legend of Noah*, chap. 6.

28. Matthew Hale, *The Primitive Origination of Mankind, considered and Examined according to the Light of Nature*, London, 1677.

29. Cecil Schneer, 'The Rise of Historical Geology in the Seventeenth Century,' *Isis*, vol. 45, pp. 256–268 (1954).

30. Waller, *Posthumous Works of Robert Hooke*, pp. 279–328.

31. Yates, *Giordano Bruno*, pp. 416–423.

32. J. A. McGuire and P. M. Rattansi, 'Newton and the Pipes of Pan', *Notes and Records of the Royal Society of London*, vol. 21, pp. 108–143 (1966).

33. Marjorie Hope Nicolson, *Mountain Gloom and Mountain Glory: The Development of the Aesthetics of the Infinite*, Ithaca, 1950. Gordon L. Davies, *The Earth in Decay. A History of British Geomorphology 1578–1878*, London, 1969: chap. 1, 2.

34. Renatus Des-Cartes, *Principia Philosophiae*, Amstelodami, 1644: sec. 188. See Robert Lenoble, 'La Géologie au milieu du XVIIe. siècle', *Les Conférences du Palais de la Découverte*, série D, no. 27, Paris, 1954.

35. See Nicolson, *Mountain Gloom and Mountain Glory*, chap. 3.

36. H. More, *Democritus Platonissans, or, an Essay upon the Infinity of Worlds out of Platonick Principles*, Cambridge, 1646: see cantos 21, 76.

37. Thomas Burnet, *Telluris Theoria Sacra: Orbis nostri Originem & Mutationes Generales, quae aut jam subiit, aut olim subiturus est, complectens. Libri duo priores, de Diluvio & Paradiso* [1680]. *Libri duo posteriores, de Conflagratione Mundi, et de Futuro Rerum Statu* [1689]. Londini. (English editions of the two parts were published in 1684 and 1690 respectively.)

38. See Nicolson, *Mountain Gloom and Mountain Glory*, chap. 5.

39. H. W. Turnbull, *Correspondence of Sir Isaac Newton*, vol. 2, Cambridge, 1960, letters 246–7.

40. John Ray, *Miscellaneous Discourses Concerning the Dissolution and Changes of the World*, London, 1692. *Three physico-theological Discourses, concerning I. The primitive Chaos, and Creation of the World. II. The General Deluge, its Causes and Effects. III. The Dissolution of the World, and Future Conflagration*, London, 1693. The essays 'Of Formed Stones' are at pp. 104–132 and pp. 127–162 respectively. See also Raven, *John Ray*, chap. 16.

41. John Woodward, *An Essay toward a Natural History of the Earth: and Terrestrial Bodies, especially Minerals: as also of the Seas, Rivers and Springs. With an Account of the Universal Deluge: and of the Effects that it had upon the Earth*, London, 1695.

42. J. A[rbuthnot], *An Examination of Dr. Woodward's account of the Deluge &c. With a comparison between Steno's philosophy and the Doctor's, in the case of marine bodies dug out of the Earth*, London, 1697.

43. Robert W. T. Gunther, *Further Correspondence of John Ray*, London, 1928: letter 214 (to Lhwyd, 1699).

44. William Whiston, *A new Theory of the Earth, from its Original, to the Consummation of all Things, wherein the Creation of the World in Six Days, the Universal Deluge, and the General Conflagration, as laid down in the Holy Scriptures, are shown to be perfectly agreeable to Reason and Philosophy*, London, 1696. John Keill, *An Examination of Dr. Burnet's Theory of the Earth. Together with some remarks on Mr. Whiston's New Theory of the Earth*, Oxford, 1698. See Nicolson, *Mountain Gloom and Mountain Glory*, chap. 6, and Davies, *The Earth in Decay*, chap. 3.

45. In a letter to Ray, 1698, printed in his *Lithophylacii Britannici Ichnographia sive Lapidorum aliorumq. Fossilium Britannicorum singulari figura insignium*, Londini et Lipsiae, 1699: pp. 128–136 His English translation of this letter is printed in R. T. Gunther, *Early Science in Oxford*, vol. XIV, Oxford, 1945, letter 200. See also [Charles King], *An Account of the Origin and Formation of Fossil Shells, &c. Wherein is Proposed a Way to Reconcile the Two Different Opinions, of those who affirm them to be the Exuviae of real Animals, and those who fancy them to be Lusus Naturae*, London, 1705.

46. *Further Correspondence*, letter 151 (to Lhwyd, 1695): see also John Beaumont, *Considerations of a Book, entituled the Theory of the Earth. Publisht some Years since by the Learned Dr. Burnet*, London, 1693.

47. *Further Correspondence*, letter 154 (to Lhwyd, 1695).

48. Johannus Scheuchzerus, *Piscium Querelae et Vindiciae*, Tiguri, 1708; Carolus Langius, *Historia Lapidum Figuratorum Helvetiae, eiusque viciniae . . .*, Venetiis, 1708.

49. Johannus Scheuchzerus, *Herbarium Diluvianum*, Tiguri, 1709.

50. Johannus Scheuchzerus, ΣΥΝΘΕΩ. *Homo Diluvii Testis et* ΘΕΟΣΚΟΠΟΣ. Tiguri, 1726.

51. Haber, *Age of the World*, pp. 107–8.

52. Balthasarus Erhardus, *De Belemnitis Suevicis Dissertatio*, Lugduni Batavorum, 1724. The biological *affinities* of belemnites continued to be disputed for many years: see the summaries of 18th-century views in M. H. Ducrotay de Blainville, *Mémoire sur les Bélemnites, considerées zoologiquement et géologiquement*, Paris, 1827, section 1re.

53. Melvin E. Jahn, 'Dr. Beringer and the Würzburg "Lügensteine" ', *Journal of the Society for the Bibliography of natural History*, vol. 4, pp. 138–146 (1963).

54. Johannus Bartholomaeus Adamus Beringer, *Lithographiae Wirceburgensis, ducentis Lapidum Figuratorum, a potiori Insectiformium, prodigiosis Imaginibus exornatae Specimen Primum*, Wirceburgi, 1726. For a modern English edition, see Melvin E. Jahn and Daniel J. Woolf, *The Lying Stones of Dr. Johann Bartholomew Adam Beringer being his Lithographiae Wirceburgensis*, Berkeley and Los Angeles, 1963.

55. Antonio Vallisneri, *De' Corpi marini, che su' Monti si trovano; della loro Origine, e dello stato del Mondo avanti il Diluvio, nel Diluvio, e dopo il Diluvio: Lettere critiche*, Venezia, 1721 (reprinted in *Opere Fisico-Mediche*, Venezia, 1733, vol. 2, pp. 305–363). Anton-Lazzaro Moro, *De Crostacei e degli altri marini Corpi che si trovano su' Monti Libri due*. Venezia, 1740. Johann Gottlob Lehmann, *Versuch einer Gerschichte des Flötzgebürgen . . .*, Berlin, 1756.

56. Godefridus Guilielmus Leibnitius, *Protogaea sive de prima facie Telluris et antiquissimae Historiae Vestigiis in ipsis naturae Monumentis Dissertatio ex schedis manuscriptis Viri illustris in lucem edita a Christiano Ludovico Scheidio*, Goettingiae, 1749.

57. Edmund Halley, 'A short Account of the Cause of the Saltness of the Ocean, and of the several lakes that emit no Rivers; with a proposal, by help thereof, to discover the Age of the World', *Philosophical Transactions*, vol. 29 (no. 344), pp. 296–300 (1715).

58. [Benoît De Maillet], *Telliamed ou Entretiens d'un Philosophe indien avec un Missionaire françois Sur la Diminution de la Mer, la Formation de la Terre, l'Origine de l'Homme, &c.* [edited by the Abbé J. B. le Mascrier], Amsterdam 1748. For a modern English edition, which distinguishes le Mascrier's alterations from de Maillet's original text, see Albert V. Carozzi, *Telliamed or Conversations Between an Indian Philosopher and a French Missionary on the Diminution of the Sea, by Benoît de Maillet*, Urbana, 1968.

59. Buffon et Daubenton, *Histoire naturelle, générale et particulière, avec la Description du Cabinet du Roy. Tome premier*, Paris, 1749.

60. Le Comte de Buffon, *Histoire naturelle, générale et particulière, Supplément, Tome cinquième*, Paris, 1778: *Des Epoques de la Nature*, pp. 1–254. For modern editions, see Jean Roger, 'Buffon, Les Époques de la Nature. Édition critique', *Mémoires du Muséum national d'Histoire naturelle*, série C, vol. 10 (1962); also Jean Piveteau, *Oeuvres philosophiques de Buffon*, Paris, 1954, pp. 117–221.

Chapter Three

Life's Revolutions

I

On 1 Pluviose in year 4 of the French Republic, or in the reckoning of the unliberated parts of Europe, on 21 January 1796, Georges Cuvier (1769–1832) read a paper *On the species of living and fossil elephants* before a public session of the National Institute of Sciences and Arts in Paris[1]. Certainly in retrospect, and even at the time, it was an occasion of outstanding importance for the history of palaeontology, because for the first time the world of science was presented with detailed and almost irrefutable evidence for the reality of extinction. The fact of extinction as a general phenomenon in the history of life, and the attempt to find a satisfactory explanation for it, dominated palaeontological discussion for the next two decades.

It was in Paris that this discussion was carried on with the greatest intensity and intellectual sophistication; and the occasion of Cuvier's paper, and indeed his own presence there, epitomise in many ways the extraordinarily vigorous scientific life of the new Republic. By a decree of the Convention, the old royal Garden and Zoo and the former royal Museum (*Jardin du Roi*, *Cabinet du Roi*), with their implicit division between the study of living and dead organisms, had been reconstituted as a single National Museum of Natural History, covering virtually all branches of science except physics and astronomy[2]. The many unequal posts in the old institutions had been replaced by twelve professorships of equal status. Among the holders of these new posts—to mention only those who were to be involved in the debate on fossils—Barthélemy Faujas de Saint-Fond (1742–1819) had been retained as professor of the still relatively new science of geology, while Jean-Baptiste Lamarck (1744–1829) had been elevated from a humble botanical post and made professor in charge of all invertebrate animals (in the classification of the time, *Insectes et Vers*). Étienne Geoffroy Saint-Hilaire (1772–1844) at the age of only twenty-one had been put in charge of vertebrate zoology, retaining mammals and birds when this vast province was divided shortly afterwards. Vertebrates were also studied, however, under the heading of anatomy; and it was to this department that Cuvier was brought in 1795, at the age of twenty-five, first as understudy to the aging Mertrud and later as full professor. During the violent phase of the Revolution, Cuvier's post as tutor to a Norman noble family had kept him in provincial seclusion; at the same time it had given him ample opportunity to follow his growing passion for natural history, and he had supplemented his earlier studies with observations on marine animals—and incidentally on fossils too. It was during his time in Normandy that he was met by a refugee *savant* from the capital, who at once recognised his exceptional talents and wrote to Paris that "at the sight of this young man I experienced the astonishment of the philosopher, who, cast upon an unknown shore, saw traced there geometrical figures". Cuvier was subsequently invited to join the staff of the National Museum, where he was to remain for the rest of his life[3].

The Museum soon became the envy of scientific Europe. German universities were in the lead, it is true, in the number of scientific professorships that they supported; and several distinguished schools of mining had already made Germany the centre of geological research. What France lacked in quantity, however, it made up in quality: nowhere else was there such a brilliant group of scientists, working in an integrated research centre covering such a large area of science, and supported so liberally by the state. Moreover, there was also the National Institute itself, which had been set up after the Revolution as an organ for the promotion and diffusion of all branches of learning, thus embodying the ideal of universal knowledge that had underlain the great *Encyclopaedia* of the Enlightenment. The former Royal Academy of Sciences had been reconstituted as one of three 'classes' within the Institute, and this 'class', misleadingly termed 'mathematics and physics', in fact embraced the study of natural history and thus complemented the activities of the Museum. In Britain, by contrast, although there was substantial teaching of natural history in Scotland[4], there was scarcely any at the English universities, and state support for science was meagre in the extreme.

At the Paris Museum, the research programme was guided by Baconian, Newtonian and Linnean ideals—in the senses that the late eighteenth century understood those three figures. The influence of Buffon, who had dominated pre-Revolutionary natural history in France, was rejected: as one spokesman for the new ideals expressed it, "the magic of the style" of Buffon's work had "retarded the progress of true knowledge in natural history, by the scorn which he had, and inspired [in others], for systems and methods", while his crowning works had been no better than "cosmological romances"[5]. What was needed was the patient collection of careful observations, the search for the simple laws of nature that must surely underlie the bewildering variety of phenomena in natural history, and the reduction of that variety into a system of rationally based classification. Into this research milieu Cuvier fitted perfectly. As curator of the Museum's anatomical material he placed great emphasis on the accumulation of the most extensive collection that could be acquired; the theoretical principles underlying his research were explicitly designed to give to comparative anatomy the precision and simplicity already attained by the physical sciences; and in substance his research was concerned primarily, as the title of his greatest work indicates, with the mapping in outline and in detail of the whole extent of *The Animal Kingdom* (1817)[6].

Although Cuvier was appointed to a post in 'anatomy', he interpreted that term in a significantly broad sense. Both anatomy and physiology, as traditionally understood, were for him sterile in isolation: they needed to be integrated into a unified study of the living and functioning organism. While in Normandy Cuvier had studied, and been profoundly influenced by, Aristotle's masterly work on the biology of marine animals; and his conception of the living organism remained throughout his life basically Aristotelian. What he termed the 'conditions of existence' of an organism were logically fundamental; and the functional coordination of the organs within the body simply expressed the material realisation of this irreducible character of living things. Organisms were essentially machine-like, and could in principle be explained in physico-chemical terms ('vitalistic' forces were abhorrent to him, because inexplicable); but like machines all the parts of an organism were integrated to produce a functional whole. The form and action of each part was connected more or less directly with the form and action of every other: no part could suffer more than trivial modification without impairing fatally the integration and therefore the continued existence of the whole. It followed that 'species' were no mere artificial abstractions from the seamless variety of nature, as Buffon had suspected, but genuine discrete units grounded in the inescapable necessities of the conditions of existence. Variation there was, indeed, within many species; but it could only affect the functionally (and often literally) superficial parts of the body, for any major variation in the essential machinery of the body would render it unworkable. Species were therefore real and stable units of the animal kingdom, each the embodiment of a distinct mode of life[7].

II

Cuvier's conception of his task as—in modern terms—the study of functional morphology received public expression very soon after his arrival in Paris, when he took over Mertrud's lecture course and styled it *comparative* anatomy. It was to be concerned with "the understanding of animal machines", and for this the most heuristic method was to study in turn the organs subserving each vital function, comparing those organs over the whole range of the animal kingdom. It would then be possible to understand fully the functional integration of all the organs within each animal[8]. Cuvier soon formulated two 'rational principles' by which to express the basis of this research. The necessary inter-dependence of all the organs of the body was manifested anatomically in the 'correlation of parts': for example, any animal that followed a carnivorous mode of life could be expected to possess not only teeth suitable for a diet of flesh but also, in correlation with that character, claws suitable for catching and holding its prey, and so on. Secondly, in trying to reduce the variety of anatomical organisation to some kind of order, Cuvier believed he could substitute for the mere empiricism of botanical classification a more 'rational' system based on the 'subordination of characters'. All vital functions were equally essential to the life of an animal; but some, in effect, were more equal than others, and the organs that served the most fundamental functions could therefore be given greater weight in deciding the natural affinities of different animals[9].

Cuvier had scarcely begun the series of detailed anatomical studies that were to justify these principles, when the National Institute received from Madrid some unpublished engravings of a giant fossil animal that had been sent there from Paraguay. Cuvier was instructed to report on them, and announced not only that the animal was new to science and almost certainly extinct—which was already suspected—but also that this rhinoceros-sized creature had belonged to the same family as the humble sloths (Fig. 3.1). Although so different in size, and no doubt in habits too, this was the surprising conclusion to which comparative anatomy led: it was "a new and powerful proof", he said, "of the invariable laws of the subordination of characters", and of the validity of those 'laws' as a means of determining natural affinities[10].

It was the study of what he termed *Megatherium* (i.e. 'huge beast') that led him to look more closely at the old problem of the other creatures of 'the ancient world' whose remains had been collected and increasingly discussed throughout the eighteenth century. The bones and teeth of elephants and rhinoceros had long been known from Siberia, but these had been fairly readily explained as the detritus of some deluge that had swept northwards from the tropical regions in which such mammals normally lived. The animal that Cuvier later named *Mastodon*, on the other hand, had posed a much tougher problem ever since its first discovery on the banks of the Ohio in 1739, for it appeared to combine elephant-like tusks with hippopotamus-like teeth. Buffon had inferred at first that they belonged to some extinct species, but his colleague Daubenton later persuaded him that they were more likely to be a fortuitous mixture of the remains of an elephant and a hippopotamus. The opinion of other naturalists wavered between these alternatives, and Buffon himself later returned to the view that a single extinct species might be represented[11]. But whether extinct or not, Buffon like other naturalists assumed that all these bones represented tropical creatures; and he regarded their presence in cold latitudes as powerful evidence for his theory of the gradual refrigeration of the globe.

Whether any species had truly been 'lost' from the world thus remained a question as uncertain and debatable at the end of Buffon's life as it had been nearly a century earlier at the end of Ray's. Many groups of fossils, such as ammonites and belemnites, were now recognised beyond all doubt as organic remains differing radically from any known living animals; but it could still be asserted with good reason

1 Paresseux didactyle ou unau

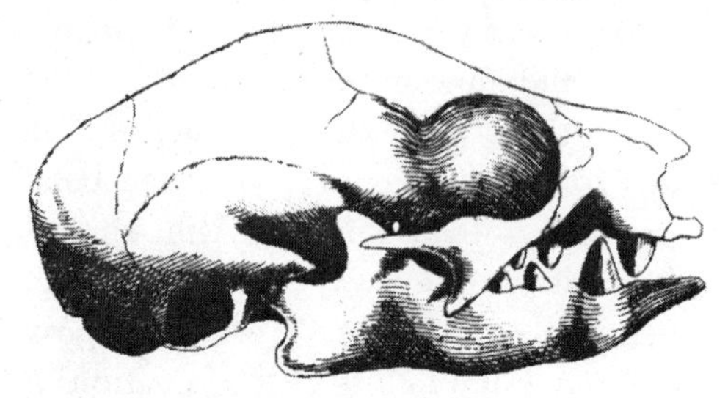

2 Paresseux tridactyle ou Ai

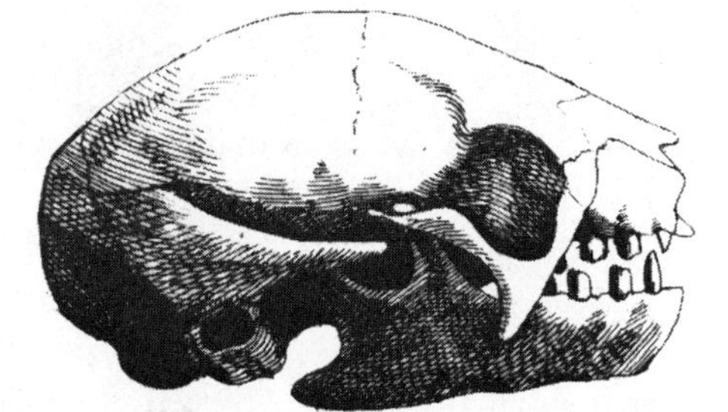

3 Animal du Paraguay

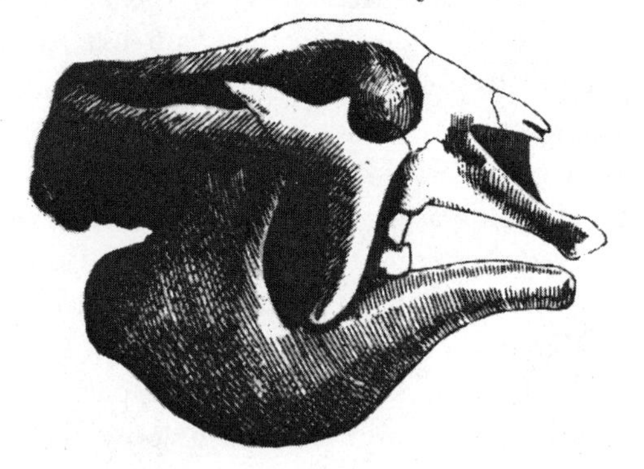

Fig. 3.1. Cuvier's first use of comparative anatomy in palaeontology: the skulls of two species of modern tree-sloth (1, 2), compared with that of the huge fossil ground-sloth Megatherium *from Paraguay (3), reduced to the same size. From a preliminary paper on the Paraguay fossil, published in 1796*[10].

that they might be living in deep water or in some remote part of the world (the discovery of the coelacanth and *Neopilina* in the present century should remind us that this argument retains its validity even today). It was Cuvier who first recognised clearly that this question, so essential for the knowledge of the history of life, would never be resolved decisively except by using the large terrestrial quadrupeds as a 'crucial experiment'. Although much of the wild interior of Africa and South America, for example, remained unexplored, it was becoming less and less likely that any *large* new mammals would be discovered alive. It was almost inconceivable that the browsing megatherium would be found alive in South America, or that the pioneers pushing westwards from the thirteen original United States of America would find themselves faced with mastodons as well as Indians. Therefore if a study of these fossil bones, using the powerful new methods of comparative anatomy, could prove that they had belonged to species distinct from any known alive, the reality of extinction would be proved almost beyond dispute.

Such a conclusion was suggested to Cuvier by the megatherium, but he knew that it would have much greater force if it could also be demonstrated on the fossil elephants that had provided so much of the material for the earlier debate. The resources of the Museum for this project had, as it happened, been enlarged shortly before by the compulsory acquisition of the Stadtholder's collection at the Hague, as a result of Republican victories in Holland. Cuvier thus had ample material with which to compare in detail, bone by bone, the skeletal anatomy of living and fossil elephants. This enabled him to assert, firstly, that the Indian and African elephants were distinct species, for their anatomical differences were too great and too constant to be merely the effects of different environments. This in itself was a triumph for the new methods in anatomy, with their emphasis on dissection and internal features. But furthermore, Cuvier was able to argue that the fossil elephant or 'mammoth' found in Siberia and northern Europe had been distinct from either of the living species (Fig. 3.2).

This was a conclusion with far-reaching implications, which probably explains why Cuvier was chosen to present it at the National Institute, where the various branches of science took it in turns to give public lectures on their latest research. Though some of these lectures were on subjects ill-adapted to a general audience, Cuvier's material and his conclusions were sensational enough to be of absorbing interest to

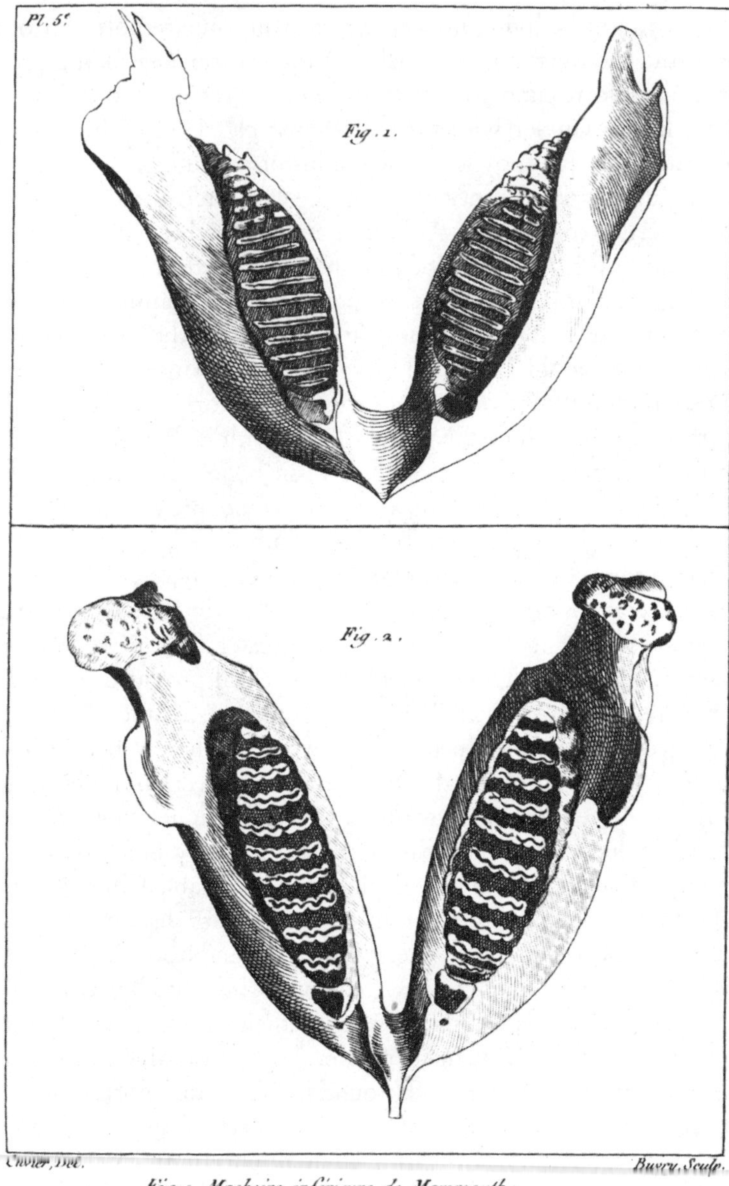

Fig. 3.2. Cuvier's first demonstration (1799) of the specific distinction between the mammoth (above) *and the Indian elephant* (below), *shown here by a comparison of their lower jaws*[12]. *Cuvier argued that this distinction implied that the mammoth was a truly extinct species.*

any *amateur* of science. Not only were the great pachyderms intrinsically fascinating as the largest of all land animals, but the demonstrable extinction of a distinct species threw unexpected light, as Cuvier said, "on the history, so piquant and so obscure, of the revolutions of this globe"[12].

III

Cuvier's use of the word 'revolution' in the context of Earth-history was not original, indeed it was a commonplace; but in earlier writers such as Buffon its overtones had been more Newtonian than political. As the planets revolved around the Sun, so in Buffon's view the Earth had undoubtedly undergone many gradual changes in the course of its long history. In a revolutionary era, however, the word took on new overtones of sudden violence, and it was in this sense that Cuvier came to view the history of the Earth as punctuated by 'revolutions'. Just as the institutions of the old régime had been suddenly swept away and replaced by new ones, so these fossil bones seemed "to prove the existence of a world anterior to ours, destroyed by some kind of catastrophe".

This was not a conclusion based merely on a fancied analogy with contemporary politics, however; it was firmly grounded, Cuvier believed, in the results of his detailed research. For if the mammoth had been a species distinct from either of the living elephants, there was no reason to suppose it had been tropical in habitat, and Buffon's explanation of its position in cold latitudes at once became suspect if not invalid. This was an inference that was confirmed some years later by a report on one of the rare mammoth carcases preserved in frozen ground in Siberia, which proved that the mammoth had had a furry skin well suited to a cold climate (a similar carcase of a woolly rhinoceros was already known). But if the mammoth had been cold-adapted, how had it become extinct? Unlike an earlier period, when the crucial question would have been 'why?', the problem was now one of 'how?': the implicit concerns that lay behind the question were no longer, at least for Cuvier, those of natural theology, but of Aristotelian functional biology. If the mammoth had been well adapted to a cold climate, and all the parts of its body had been functionally integrated to serve a definite mode of life in that climate, what could have caused it to become extinct? Any gradual change in its environment could surely have been met simply by migration to a more suitable area (as indeed Buffon had suggested in his *Epochs*): only some sudden and drastic event, in Cuvier's view, could have overwhelmed a manifestly successful species so completely.

It was not only his fundamental conception of organic life, however, that led him to assume that "some kind of catastrophe" must have been responsible. In addition one of the most respected of contemporary geologists had assembled a formidable body of evidence to suggest that some such event had indeed occurred in the relatively recent past. The Genevan naturalist Jean-André De Luc (1727–1817), in a long series of diffuse publications[13], had argued in detail that many of the geological processes at present in operation could not be traced very far into the past: for example the Rhône was steadily building a delta out into the Lake of Geneva, but could not have been doing so for an indefinite period of time or the whole lake would long since have been reduced to an alluvial plain. In the light of much later research we can see that De Luc had good grounds for maintaining that there was some kind of discontinuity only a few thousand years back in Earth-history: all over Europe the end of the Pleistocene Ice Age had indeed produced radical changes in the nature and rate of geological processes. It was an awareness of such phenomena that made it difficult for most geologists at the time to accept the arguments put forward by the Scottish philosopher James Hutton (1726–97) for the uniformly slow action of *all* geological processes. Cuvier believed that the difference between such men as Saussure, Pallas and Dolomieu on the one hand, and such system-builders as Buffon on the other, had lain in the determination of the former to adhere to the principle of actualism:* "all System [that is grandiose speculation] has been rejected by them", he said approvingly: "they have recognised that the first step to make in order to find out about the past, is to understand clearly about the present"[15]. However it could not be assumed *a priori* that the present was in all respects an adequate or fully representative sample of the past: some natural events might occur so infrequently,

* The term 'actualism' is used throughout the present book to denote the methodology of inferring the nature of past events by analogy with processes observable in action at the present: this term, derived from Continental usage (*actualisme, Aktualismus*), is preferable to the term 'uniformitarianism' more commonly in use among English-speaking geologists, since the latter was originally intended to emphasise not the methodology but the *content* of a particular theory, *viz.* Charles Lyell's, as a 'steady-state' system of Earth-history[14].

relative to the short period of human records, that the method of actualism would have to be replaced by direct inference from the preserved effects of such events. It was some such exceptional event, of a kind unrecorded in more recent human history, that seemed to be implied in the varied phenomena that De Luc described.

De Luc's own reasons for wishing to prove the recent occurrence of a drastic physical event were explicitly related to his desire to show that the historicity of the Flood recorded in scripture was attested by scientific evidence. Like earlier theorists with the same concern, he was obliged in practice to adopt a very flexible interpretation of the Flood, and he conceived it as an engulfment of the pre-diluvial continents by crustal collapse, followed by the emergence of the present continents from what had been the floor of the pre-diluvial oceans. Although De Luc still used the old arguments from chronology to help to establish the recent date of the Flood itself, he was prepared to concede an indefinitely long period of geological time before this event, adopting what had long been a standard interpretation of the 'days' of Creation as eras of unstated length.

This concern to reconcile geological evidence with scripture, and specifically with the Flood-narrative, had ceased to be of much consequence in scientific circles generally. Only in England, backward still in this as in other scientific matters, did such efforts continue to worry men of science, and still more, the general public, for many years more[16]; and significantly it was in England that De Luc made his home for most of his life. However De Luc's old-fashioned concern with scripture did not detract from the scientific validity of his detailed arguments for the occurrence of *some* kind of drastic physical event in the geologically recent past. Cuvier, like his older colleague Dolomieu, was thus able to borrow De Luc's concept of a recent 'revolution' without having to accept with it De Luc's attempted concordance with scripture[17]. In Cuvier's opinion most of the available ancient literatures—including the Old Testament as one among many—preserved some more or less garbled tradition of an ancient Flood-event. But much more important and reliable evidence on the nature of the 'revolution' would be found, he believed, by a careful study of its *natural* effects.

Thus in the first place the revolution that had destroyed the mammoth seemed likely to have been the event that De Luc described, because the fossil mammoth bones were found in the superficial gravel deposits that De Luc dated from this episode. Such deposits were evidently recent, at least in geological terms, for they overlay irregularly the ordinary series of strata and were often confined to the valleys of present rivers, excavated in those strata. By the time Cuvier published the full version of his memoir on elephants (1806)[18], he had found some bones with oysters and other marine organisms attached to them, which seemed to confirm that the 'revolution' had been some kind of prolonged marine incursion, and that it had not been a mere transitory flood. Furthermore, since the bone-bearing gravels were confined to fairly low-lying areas, he concluded that this incursion had not been universal but merely local. Finally, since the bones themselves were well preserved and showed few signs of abrasion, he inferred that they had not been transported from a distant region and that they were the remains of animals that had lived and died near where they were found. He therefore concluded that the 'revolution' that had caused the extinction of the mammoth had been a sudden but prolonged inundation of low-lying areas by the sea. Alternatively, being greatly impressed by the occasional preservation of carcases of these species in the permafrost areas of Siberia, he thought it might have been a sudden drop in temperature. He never clearly resolved the conflict between these two hypotheses, and still less was he prepared to speculate on the physical cause of the event, though he evidently believed the cause was natural. But this was simply a product of his characteristic scientific caution. His own work, he believed, had by its rigorously factual basis proved the inadequacy of earlier hypothetical 'systems' of Earth-history, and especially Buffon's: he was therefore most concerned to avoid the methodological error of his predecessors in allowing speculation to outstrip observation.

IV

Cuvier's research in the few years following his first lecture—some of the most productive years of his life—proved conclusively that it was not just the mammoth and the megatherium, but a whole fauna that had disappeared. In one detailed memoir after another, each founded on the most painstaking comparison with the anatomy of living mammals, he demonstrated that there had been species of elephant, hippopotamus, rhinoceros, armadillo, deer, cattle and so on—all distinct from any living species, many of them much larger in size, and all apparently vanished from the face of the Earth. The resurrection of such a spectacular zoo was a striking accomplishment indeed, and it brought Cuvier fame throughout the world of science.

To Cuvier himself, however, though he was never a man to shrink modestly from the limelight, this work was equally satisfying as a vindication of his biological principles. Each fossil animal that he reconstructed was a monument to the principle of the coordination of parts, and its assignment to its appropriate place in the animal kingdom was likewise a demonstration of the validity of the principle of the subordination of characters. Most fossil quadrupeds, unlike the complete megatherium and the occasional mammoth, were found only as scattered and disarticulated bones, and often the bones of several species were found mixed in the same deposit. There was thus a risk that an imaginary composite animal might be assembled in error, just as Daubenton had suspected in the case of the remains from the Ohio. But by the application of his two anatomical principles Cuvier believed that this risk could be avoided, and that each species could be reconstructed with complete reliability (Fig. 3.3). If the form of each bone were analysed in functional terms, in relation to the rest of the body, it would be possible to piece together the functionally integrated construction of the whole animal, and there would be no risk (to take a crude example) of fitting a carnivore's jaws on to a skeleton with a herbivore's feet. The correlation of parts thus became, with fossil material, a heuristic principle with predictive value: its application indicated which bones in a mixed collection had belonged to which others. In Cuvier's view it had the supreme virtue of being a *rational* principle; but in practice, since the functional significance of many anatomical correlations was far from certain, he was often forced to rely on the *empirical* observation that certain features were in fact generally associated in a given group of animals. In practice, therefore, the predictive value of his comparative anatomy rested more on his extremely wide knowledge of living animals than on the principle of functional coordination that he prized so much. When, for example, he discovered a fossil of marsupial appearance in the gypsum quarries of Montmartre, he believed that his principles would be vindicated if further excavation of the specimen were to reveal that it had possessed marsupial bones in the pelvis. The prediction was duly confirmed, and this result was claimed by Cuvier

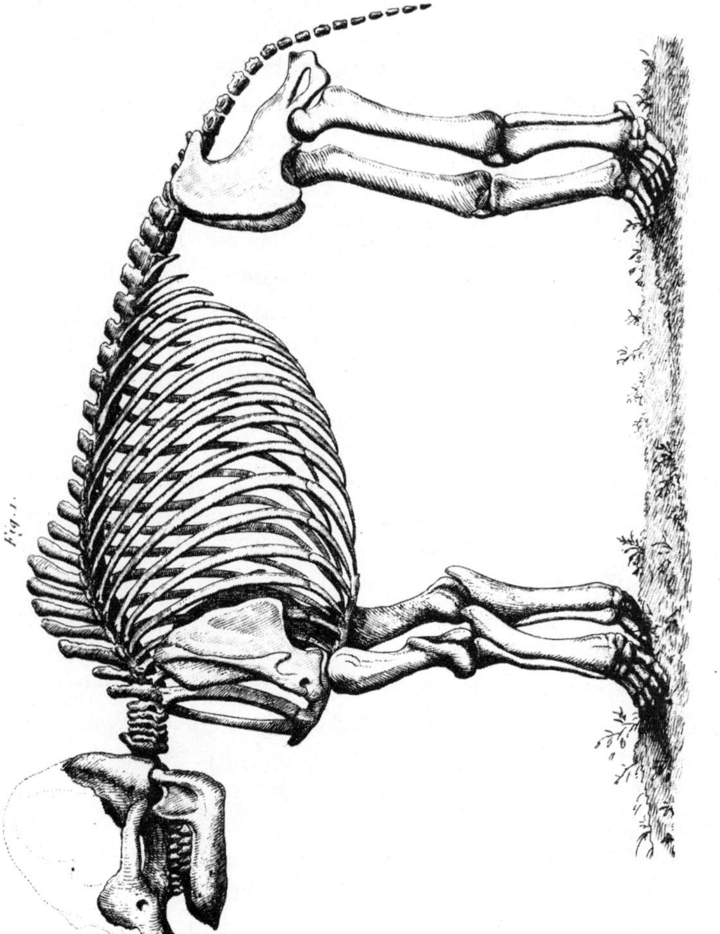

Fig. 3.3. Cuvier's reconstruction (1806) of the American fossil Mastodon[13]. *This was a spectacularly successful result of the application of his anatomical methods, for earlier naturalists had suspected that the scattered bones and teeth of this mammal might have belonged to more than one species.*

as a spectacular demonstration of the truly scientific status of his principles (Fig. 3.4); but in fact it was grounded more in a comparison with the living marsupials of America and Australia than in a functional correlation with the marsupial pouch[19].

Nevertheless, whatever the real logical status of his principles, there is no doubt that Cuvier's work was strikingly successful in the eyes of his contemporaries, and seemed to them to justify his hope that anatomy would soon be expressed in terms of 'laws' as simple and, as it were, as 'mathematical' as those established by Newton in physics and now by the new chemistry of Lavoisier. The scientific prestige of Cuvier's comparative anatomy, combined with the immediately striking nature of his reconstructions of fossil mammals, thus assured that his wider speculations would be received with equal respect.

Five years after his earlier lecture to the Institute, Cuvier was able to summarise the fruits of his research and draw out their implications more fully than before in a lecture 'On the species of quadrupeds of which the bones are found in the interior of the Earth' (1801)[15]. It was now clear that the systematic study of fossils by the methods of comparative anatomy could yield decisive evidence on which to base a 'theory of the Earth' more firmly than had been possible in the days of Buffon. Recent research had demonstrated beyond doubt that in the relatively recent past the Earth had been inhabited by animals at present unknown alive. Therefore the most important question, Cuvier pointed out, was "to discover if the species which existed then have been entirely destroyed, or if they have merely been modified in their form, or if they have simply been transported from one climate into another". Here, clearly stated, were three alternative explanations: extinction, evolution or migration. The last of these, Cuvier admitted, could still be regarded as a valid possibility for marine animals, but for the larger terrestrial quadrupeds it could now be disregarded. That, in Cuvier's view, left extinction *versus* evolution as the effective choice.

V

Such a stark antithesis may seem in retrospect peculiar, because Darwinian theory was later to integrate the two by making extinction an important aspect of the mechanism of evolution itself. But for

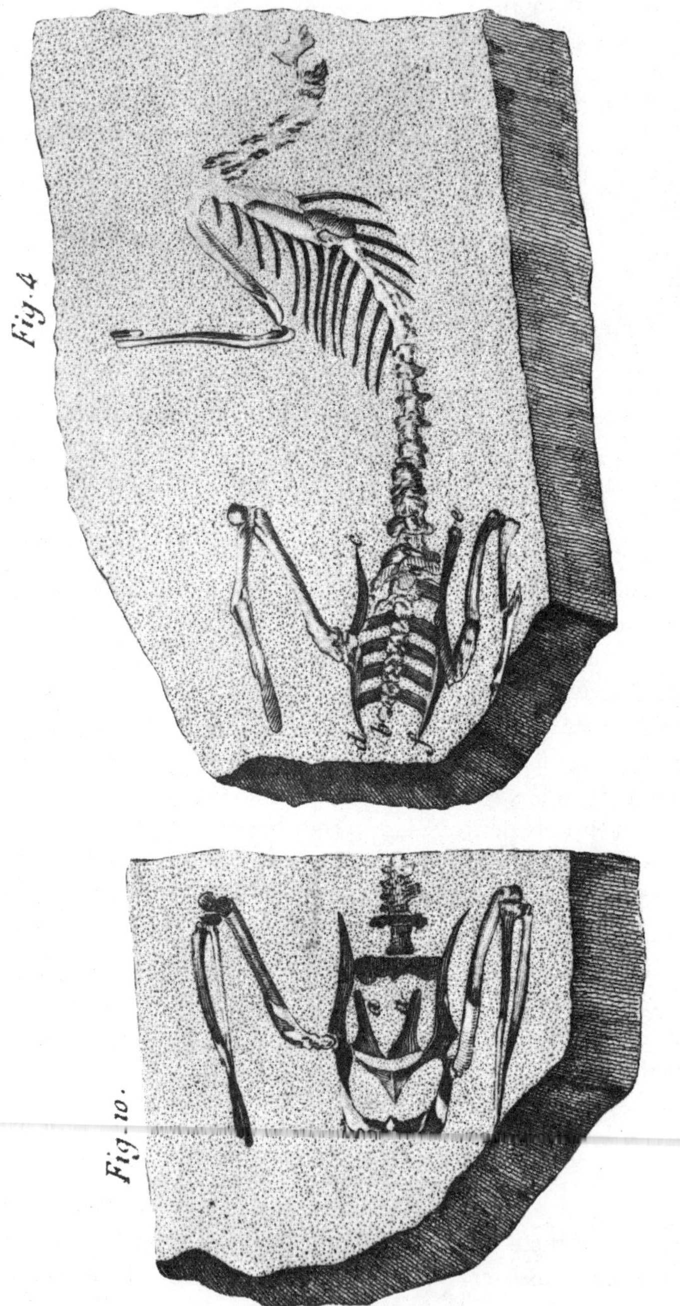

Fig. 3.4. *Cuvier's drawings (1804) of an incomplete skeleton of a fossil marsupial from the early Tertiary gypsum beds of* Montmartre. *His dissection of this specimen (right) revealed the marsupial bones (left, a, a) that he had predicted in accordance with his anatomical principles*[19].

Cuvier the suggestion that fossil animals had "merely been modified in their form" to become living species was a theory that virtually denied the reality of extinction altogether. This theory, championed particularly by Cuvier's older colleague Lamarck, although in retrospect labelled 'evolutionary', has little resemblance to Darwin's later theory of evolution. It is therefore important to note that Cuvier's rejection of evolution—for which he was formerly branded as a reactionary villain in the history of science—was primarily a defence of *extinction*, not of special creation. The question at issue at this stage was not the origin of present species but the fate of former ones.

Lamarck not only belonged to an older generation than Cuvier but also adhered to an older scientific tradition, and his evolutionary theory can only be understood fully as part of a much wider philosophy of nature. He believed not simply that one biological species evolved into another, but rather that species themselves were unreal: they were arbitrary and artificial divisions within the seamless variety of nature. The word 'species' still retained for him its original breadth of meaning, and such divisions were as unreal among, for example, chemical substances and minerals as they were among animals and plants[21]. However, while the diversity of organisms was in a sense seamless, it was not therefore formless. The true order and system of organic nature was indicated, Lamarck believed, by the ancient concept of the 'scale of beings', on which the different groups of organisms could be arranged according to the 'perfection' of their faculties. He was sufficiently impressed by the fundamental differences between animals and plants to recognise that these two kingdoms of nature could not be arranged serially in a single scale; and since he also emphasised strongly the contrasts between living and non-living entities he did not attempt to extend either scale 'downwards' into the mineral kingdom. Nevertheless, for either animals or plants the concept of 'higher' and 'lower' forms was for him no mere metaphor but a reflection of the true order of nature.

In one form or another the scale of beings had been a commonplace in eighteenth-century natural history[22]: it was admirably adapted to the ideology of the Enlightenment, since it made Man, at the top of the scale, literally the measure of all other beings. Its terminology sometimes has an anachronistic and spurious similarity to the language of later evolutionary theories; but generally it was a static concept of a hierarchy that had not changed or developed in the course of time.

It was only after he began to work on invertebrate animals that Lamarck himself seems to have begun to transform the scale from a static to a temporal concept. His transfer from botany to zoology when the Museum was founded reflects his desire to find in the 'lower' animals a new field for demonstrating the insensibly gradual transitions of form within organic nature[23]. But by 1800, when he introduced a course of lectures at the Museum, he stressed the importance of the invertebrates not only as "the most imperfect animals, the most simply organised", but also as the most primitive in a temporal sense. They were, he said, "perhaps the ones with which nature began, while it formed all the others with the help of much time and of favourable circumstances"[24].

However, this change of opinion was less radical than it might seem: it merely meant that Lamarck had become convinced that animal 'species' were as unreal as those of minerals, and that all natural entities were involved in the same process of continual flux. He had in fact been devoting much of his time at the Museum to the study of geology, not zoology. In his work entitled *Hydrogeology* (1802)[25] he suggested that the supposedly 'primitive' rock granite must have been transformed gradually from organic debris, "given time and favourable circumstances"—the same phrase that he used of organic species. With such speculations, flying in the face of Lavoisier's new chemistry, it is not surprising that Lamarck's book was unable even to find a commercial publisher; he was forced to publish it at his own expense and, as he complained bitterly, it was virtually ignored by his colleagues. However it is an important clue to Lamarck's evolutionary thought, not least because it supported the immensely long time-scale that his theory required. As the title suggests, it was a geological system that gave no place to igneous processes within the Earth: as in Buffon's earlier work (though not his later *Epochs*) it envisaged a series of slow changes in physical geography taking place solely by the aqueous agencies of erosion and deposition. Lamarck had little understanding of stratification, and simply used the widespread occurrence of fossils on land as an argument for a slow inexorable interchange in the positions of the continents and oceans. As a result of the generally westward trend of oceanic currents, the continents were steadily being eroded on their western shores while gaining land by silting on their eastern: given a long enough time the continental masses would thus describe a complete traverse of the globe, and indeed Lamarck assumed that they had already encircled the Earth several times in this way.

In its steady-state concept of Earth-history, its virtually eternalistic time-scale, and its marginal use of fossil evidence, Lamarck's theory has obvious affinities with James Hutton's more famous *Theory of the Earth* (1788)[26]. Although they differed in the mechanisms they proposed for ensuring a steady state, the motives for propounding such theories were the same: within the framework of deistic natural theology only a steady state could demonstrate the wisdom of nature. As in an earlier period, the vast magnitude of time postulated in such theories was not in itself a barrier to their acceptance. Most geologists in the late eighteenth century were justifiably reluctant to commit themselves to quantitative estimates, since they had no firm evidence to go on; but it is clear, from frequent comments on the thickness of evenly deposited strata, that they realised that geological time had been extremely lengthy by human standards. When, for example, the veteran geologist Nicholas Desmarest (1725–1815) discussed Hutton's theory, breaking his rule of excluding living authors from his survey of geological theories in recognition of Hutton's importance, he criticised the theory not for its vast time-scale but because its cyclicity was insufficiently grounded in detailed observations[27]. It is not surprising that Lamarck's theory was virtually ignored, for by 1802 the progress of empirical knowledge of geology was rightly felt to have made such speculative theories inexcusable.

Whether scientifically excusable or not, however, it was this theory of virtually infinite time that gave Lamarck the justification he needed for interpreting the phenomena of fossils in evolutionary terms. In his great *System of Invertebrate Animals* (1802) he asserted that "one must believe that every living thing whatsoever must change insensibly in its organisation and in its form"; and given enough time such slow changes could have been sufficient to produce the whole observed range of animals and plants. The differences between living and fossil species were not in the least surprising, he argued, for there had been ample time for the fossil forms to change into the living: on the contrary, what was surprising was to find any living species whatever in the fossil state, and it could only be inferred that these were so relatively recent that they had not yet had time to change. "One must therefore never expect to find among living species all those which are found in the fossil state", he concluded, "and yet one may not assume that any species has really been lost or rendered extinct"[28].

Lamarck's evolutionary theory thus excluded the possibility of extinction, and was therefore inevitably opposed to Cuvier's conclusions. But the disagreement ran deeper than that. The two naturalists held diametrically opposed views of the nature of organic life, and it was this that made their dispute so vehement. To Cuvier, as we have seen, the organism was a functionally stable mechanism, and the species therefore a temporally stable unit of nature. How such units had come into existence was for him a matter on which no scientific evidence was available and on which it was therefore no business of the scientist to speculate. Once in existence, however, the processes of generation would ensure the maintenance of the organisation appropriate to the species' mode of life, and only some dramatic environmental change could eliminate it from the face of the Earth. To Lamarck, on the other hand, such 'mechanising' of biology was as abhorrent as the changes that Lavoisier had effected in chemistry. Life was essentially a process of continual flux, species were unreal, and it was inconceivable that the insensible gradations of the scale of animals and plants could ever have been left imperfect by gaps caused by extinction. Cuvier objected to this attitude not because it threatened some assumption of special creation (a doctrine he never expounded) but because Lamarck's beliefs would in his opinion make the study of organisms unscientific and pointless: "In a word, it would be to reduce to nothing the whole of natural history, since its subject would consist only of variable forms and fleeting types"[29].

VI

The clash of opinion between Cuvier and Lamarck as to the interpretation of fossil species received almost at once a piece of important empirical evidence; but like most 'crucial experiments' in science it appeared crucial only to one side. Among the *savants* who accompanied the French military expedition to Egypt the Museum was represented by Geoffroy, who made it his business to assemble a collection of mummified animals from the ancient tombs. Back in Paris these specimens caused almost as much excitement in the world of learning as that more famous trophy of the same expedition, the Rosetta Stone. For here at last was a chance, as the Museum stated in its report, to determine "whether species change their form in the course of time".

It was recognised, of course, that even the most recent fossils had an "incomparably more remote origin" than these mummified specimens; but nevertheless, if, as Lamarck asserted, *time* was essential to the process of change, it was legitimate to see whether a short but known time-span of a few thousand years had in fact effected at least a small change in the form of organisms[30]. Methodologically this was no different from the procedures of the prestige science of astronomy, in calculating (say) the lengthy orbits of comets or the outer planets from observations of their movements over a shorter period of time. It is significant that it was an astronomer, John Playfair (1748–1819), who was at this very time arguing for the use of the same quasi-mathematical reasoning in geology, in his *Illustrations of the Huttonian Theory of the Earth* (1802)[31], applying the idea of an 'integration' of small changes to such problems as the gradual excavation of valleys.

The result of the examination of the mummified animals at the Museum was unambiguous: each species was indistinguishable from living specimens, not only in its osteology but also in the comparative anatomy of all the organs preserved by the enbalming process. The sacred ibis (Fig. 3.5) might have been an exception, but Cuvier in a special memoir showed that its differences from the living bird of that name were due to mis-identification[32]. The crucial experiment therefore seemed wholly in favour of Cuvier's conception of organic stability, and contrary to Lamarck's conception of organic flux.

But Lamarck, although he signed the Museum's report on the Egyptian collection, was unconvinced by the experiment. Introducing a monograph on the fossil molluscs from the strata around Paris, he argued that the modifications of so-called 'species' were unobserved simply because they took place so slowly relative to Man's time-scale: to conclude that species were stable was as fallacious, he suggested, as it would be for insects living in a building, and seeing no change over twenty-five of their own generations, to conclude that the building was eternally stable[33].

This might suggest that the grounds of the dispute between Lamarck and Cuvier lay in a difference in the time-scales that they envisaged. But in fact Cuvier was prepared to suggest an age of "some thousands of centuries" for the fossil-bearing strata around Paris, which was about as high an antiquity as a disciplined scientific imagination

Fig. 3.5. Cuvier's drawing (1804) of the skeleton of the Ibis from ancient Egyptian mummified specimens[32]*. He successfully identified this with a living species of bird, and proved that it had undergone no 'transmutation' (that is, evolutionary change) in the last three thousand years.*

could properly have inferred from the evidence.* Lamarck probably envisaged an even longer time-scale, but in any case his dispute with Cuvier lay more in their disagreement about the proper methods of inference to be adopted in a 'historical' science like geology. Lamarck, believing "that time and favourable circumstances are the principal means that nature uses in creating all its products", was prepared to conclude that as "we know that time has no end to it and is consequently always at its [that is, nature's] disposal", the transmutation of species was self-evidently true. To Cuvier, on the other hand, such metaphysical apriorism was utterly subversive of any true science. "I know", he wrote later, "that some naturalists rely greatly on the thousands of centuries which they pile up with a stroke of the pen; but in such matters we can scarcely judge what a long time would produce, except by multiplying in thought what a lesser time produces"[44]. Such a lesser time, although admittedly short by geological standards, could be shown to have produced no detectable changes whatever; and therefore, in Cuvier's view, Lamarck's concept of organic flux was without scientific foundation.

Lamarck indeed had no positive evidence for evolution from the fossil record, beyond the fact that the fossil molluscs around Paris were of different species from any known alive; and he tended to give a different explanation even for that fact, for he used the tropical character of some of the genera (for example *Nautilus*, *Murex* and *Cerithium*) as an argument for a highly speculative theory of slow polar wandering[34]. It is therefore not surprising that when he wrote a major exposition of his evolutionary theory, entitled *Zoological Philosophy* (1809), fossils were scarcely mentioned at all. The work elaborated a general theory of the temporal actualisation of the scale of animal beings, but it interpreted this gradual ascent of the scale only in terms of an inherent 'tendency' of life to improve itself in the course of time. This unexplained 'power' was now, in Lamarck's system, the main 'cause' of evolution. The action of 'favourable circumstances' (the 'essential' Lamarckism of later biologists) was only a secondary explanation introduced to account for the fact that, owing to their adaptations to different modes of life, organisms could not in practice be arranged on a linear scale with insensible gradations[35].

* In modern terms the strata, of Eocene, that is early Cainozoic, age, would be dated at roughly fifty million years; but even in the late nineteenth century geologists were quite willing to agree with physicists on a figure much lower than this, and hence not far from what Cuvier probably had in mind.

VII

Ironically, at the very time that Lamarck was composing his evolutionary theory without any fossil evidence for the progression of life in the course of time, Cuvier's work was beginning to uncover the first signs of just that class of evidence. It was this research, greatly enlarged and extended by later palaeontologists working in the Cuvierian tradition, that ultimately helped to make Darwin's theory scientifically convincing where Lamarck's was not.

In his early research, Cuvier was clearly thinking in terms of a single revolution and a single "world anterior to ours". This was natural, because the easiest fossil bones to tackle had been those from the superficial deposits of gravel; and it was from these that most of his extinct fauna had come. However, unlike De Luc, Cuvier did not mean to attribute any unique significance to this recent 'revolution'. Even in his first lecture at the Institute, he suggested—to an audience all too well aware of continuing political instability—that just as the animals he had reconstructed had been annihilated and replaced by those at present in existence, so these latter in turn would "perhaps be likewise destroyed one day and replaced by others". Before long, however, his further research confirmed this picture of an Earth-history punctuated by occasional revolutions, by giving him evidence of more than one such event in the past.

This new dimension to his research, which was to have far-reaching consequences for the reconstruction of the whole history of life, emerged from his studies of fossils from the gypsum quarries at Montmartre. What he had identified tentatively as a fossil dog proved, on closer application of his anatomical principles, to be three distinct species of a totally unknown genus, combining some of the characters of a tapir, a rhinoceros and a pig, and thus further removed from living animals than any fossil he had studied hitherto (Fig. 3.6)[36]. Furthermore, having at first discussed the Montmartre fossils as though they were more or less of the same age as the gravel bones, he now recognised that they must be far more ancient, for the gypsum beds were part of a thick series of regular strata occupying a wide area around Paris. It was at this point that Cuvier's work, hitherto essentially anatomical, became firmly linked with the mainstream of contemporary geological research.

The development of stratigraphy, as it came to be termed, had

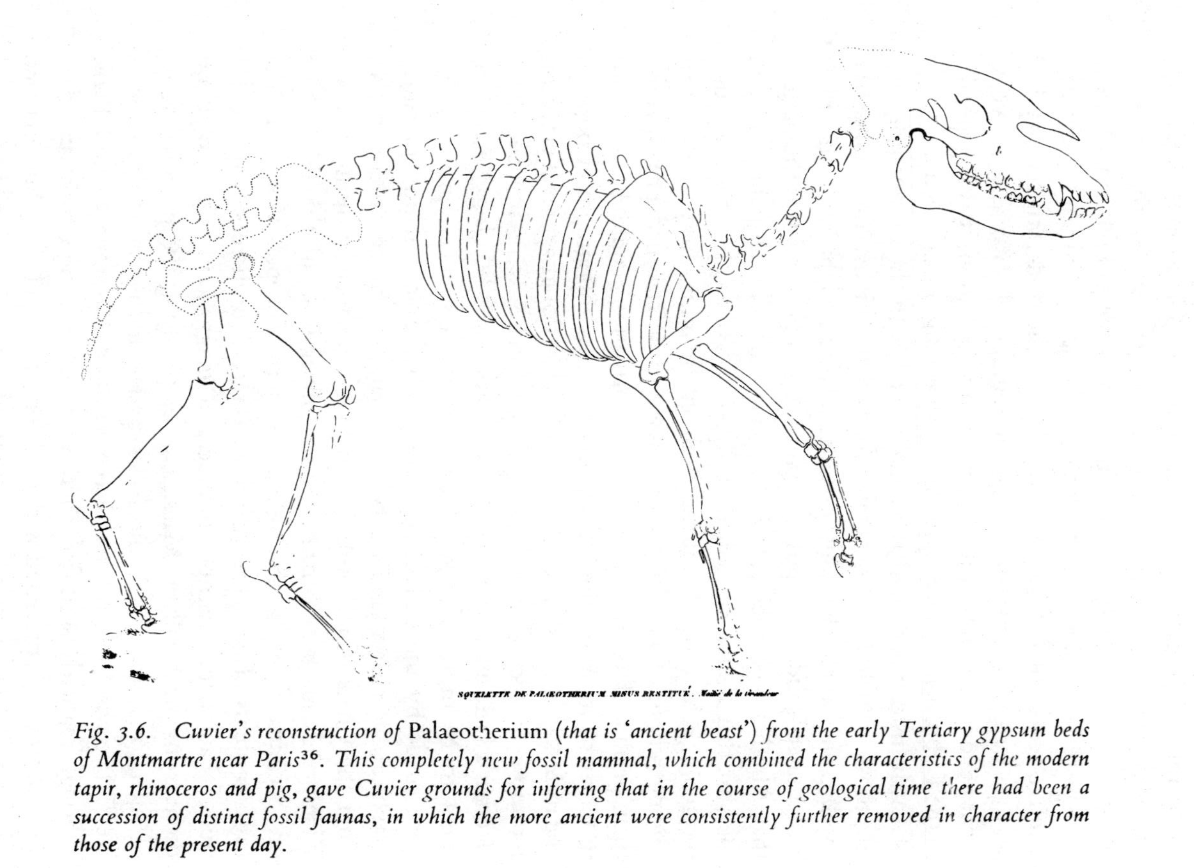

Fig. 3.6. *Cuvier's reconstruction of* Palaeotherium *(that is 'ancient beast') from the early Tertiary gypsum beds of Montmartre near Paris*[36]. *This completely new fossil mammal, which combined the characteristics of the modern tapir, rhinoceros and pig, gave Cuvier grounds for inferring that in the course of geological time there had been a succession of distinct fossil faunas, in which the more ancient were consistently further removed in character from those of the present day.*

been centred in the later eighteenth century on Thuringia and Saxony, where the strong economic incentives of an important mining industry had encouraged the detailed description and mapping of the succession of strata. The strata themselves (in modern terms of Permo-Triassic age) were conveniently distinctive in character; and the classic descriptive works of Lehmann and Füchsel in the mid-18th century had set a pattern for many other local monographs. Insofar as such works went beyond description to theoretical interpretation, it was generally in the Leibnizian terms of a gradually shrinking ocean. Such a concept of Earth-history accorded moderately well with the observation that the most ancient rocks generally formed the higher mountains, the fossiliferous strata the hills of intermediate height, and the most recent unconsolidated deposits the lowest ground. This was the pattern that Abraham Werner (1749–1817) had naturally adopted when, in the year that Buffon published his *Epochs of Nature*, he added a course on geology to those he already offered in mineralogy at the great mining school of Freiberg. Rocks were to be classified primarily by their period of origin, though each 'formation' could be described in terms of its mineralogical composition, its topographical position and its fossil contents[37]. Fossils, however, were not given great emphasis by most of the distinguished geologists who were influenced by Werner's brilliant teaching, because position and lithology seemed to them to be more reliable guides. Nevertheless, much progress had been made by the end of the century in the compilation of stratigraphical successions and in their correlation between different areas. Thus the Chalk, for example, formed a conveniently distinctive and widespread 'marker-horizon' for the upper limit of the *Flötzgebirge* or 'Secondary' strata throughout western Europe. Moreover, the fossil contents of each formation were receiving increasingly detailed study: Faujas for example produced a magnificent monograph on those from the Chalk of Maestricht, emphasising the potential value of such local studies for the fuller understanding of the history of life[38]. Likewise it had been recognised that the strata around Paris, although regularly stratified, were younger than the Chalk; and they were commonly correlated with the 'Tertiary' strata of the Subappenine hills in Italy. The tranquil deposition of such strata clearly pre-dated the excavation of the present valleys, which in turn contained the bone-bearing gravels that witnessed to the most recent revolution.

Thus Cuvier must have been made aware, perhaps by his geologist colleague Faujas, that his strange 'ancient beast' (*Palaeotherium*) was far older than his fauna of mammoths and rhinoceros; while even the Montmartre fossils in turn were recent by comparison with the huge crocodile-like animal that Faujas himself had described from the Chalk (Fig. 3.7). In his lecture to the Institute in 1801, Cuvier was therefore able to say that "the most remarkable and most astounding result" of his research was that "the older the beds in which these bones are found, the more they differ from those of animals we know today". In itself, this idea of a gradual approximation towards the present fauna was more of a commonplace in geology than Cuvier seems to have realised: it was well known, for example, that fossils such as ammonites and belemnites, which were generally suspected of being extinct, were confined to Secondary strata, while the Tertiary strata were characterised by molluscs (such as those that Lamarck was describing from the Paris area) broadly similar to those of present seas. However in view of the continuing uncertainty about the possible survival of the apparently extinct marine invertebrates, Cuvier was justified in regarding the results of his own research on terrestrial quadrupeds as a new advance in knowledge, for they established the principle of faunal change with much greater certainty.

At the same time, however, Cuvier's recognition of this principle raised the further problem of the causation of these changes. It was soon evident that a whole extinct fauna was represented in the Montmartre rocks, comparable in diversity to the extinct fauna of the superficial gravels. If this fauna had been destroyed by some earlier revolution similar to that which had annihilated the mammoths, the clues to that earlier event would surely be found by a more detailed study of the Parisian strata. For this task Cuvier joined his friend and contemporary Alexandre Brongniart (1770–1847) in a series of traverses across the region around Paris, which enabled them to describe the stratigraphical context of Cuvier's fossils. Superficially their *Essay on the mineral geography of the Paris region* (1808)[39], with its straightforward description of the successive strata from the Chalk upwards, appears to differ little from other stratigraphical monographs of the period (the term 'mineral geography', like Werner's term 'geognosy', was used to imply a factual description of rocks, as opposed to the more interpretative overtones of 'geology'). But their memoir incorporated two innovations of great importance.

Fig. 3.7. The discovery of the jaw of a gigantic fossil reptile in underground workings at Maestricht. This dramatic scene, which is taken from Faujas de St Fond's monograph on the Chalk fossils of Maestricht (1799)[38]*, epitomises the contemporary excitement at the recovery and reconstruction of spectacular extinct species. (Faujas thought this fossil was crocodilian; Cuvier soon afterwards proved by comparative anatomy that its affinities were with the lizards, and termed it a mosasaur.)*

They distinguished seven major formations above the Chalk, easily separated from one another by their general lithology. However within even a single one of these, the *Calcaire grossier* ('coarse limestone') from which most of Lamarck's fossil molluscs came, they found a constant order of superposition of individual strata over distances of more than 120 km. This remarkable constancy they were able to recognise by noting the precise nature of the fossil species in each stratum: it was, they said, "a mark of recognition which up to the present time has not deceived us". The fossils of successive strata were not wholly distinct; but those characteristic of one stratum tended to be less abundant in the next, and to be gradually replaced by a different set of species. What they had discovered, in effect, was that fossils could be used not merely to characterise in general terms a whole formation of strata, but to identify in much greater detail the individual strata within a formation.

In a more general sense, it was already well known that successive formations could be broadly characterised by their fossil contents. For example, Freiesleben's fine monograph on the stratigraphy of Thuringia (1807–15) mentioned as a matter of course the general character of the fossils of such formations as the *Muschelkalk*, along with their lithology, and cited references to the appropriate monographs where more detailed descriptions of the species could be found[40].

In England the civil engineer William Smith (1769–1839)—Cuvier's almost exact contemporary—had independently noted several years earlier that the successive Secondary formations in England were characterised by different species of invertebrate fossils[41]. However his lack of education, relatively low social position, and lack of leisure from practical work had all prevented him from publishing his vast empirical knowledge of these matters in a form that would have enabled the international scientific community to judge its worth. Cuvier and Brongniart may possibly have seen a brief notice of Smith's work; but even if they had, they would simply have regarded their own work as confirming the principle of correlation by fossils for Tertiary strata with as great precision and detail as Smith had demonstrated it for Secondary strata. A priority dispute with nationalistic overtones developed later between the French and English stratigraphers; but whatever the priorities, the fact is that in the years after Cuvier and Brongniart's memoir was published much greater attention was given to the fossil contents of individual strata, and correlation by fossils was found to be a principle of great practical value. It remained, however, an essentially *empirical* principle; and since a more biological approach to fossils tended at the same time to emphasise the ecological implications of characteristic faunas, there was room for debate on the extent to which fossils could also be regarded as characteristic of the relative ages of strata.

It was this more biological approach that gave Cuvier and Brongniart's work its second important feature. Using the actualistic principle of comparison with the present, they demonstrated that the Tertiary formations around Paris represented an alternation between freshwater and marine conditions. While some of the formations contained molluscs of genera now found only in the sea, others contained genera of exclusively freshwater habitat. The immediate significance of this conclusion was that it threw doubt on the conventional 'directional' model of Earth-history, which envisaged the gradual emergence of the continents from beneath a gradually shrinking ocean. It suggested instead some kind of rhythmic or cyclic alternation of depression and elevation of the continental areas; and it was therefore rightly seen as a possible confirmation of the kind of steady-state (that is non-directional) Earth-history that Hutton had more speculatively proposed.

However Cuvier and Brongniart did not follow Hutton in assuming that such movements must have been insensibly gradual. (In fact, Hutton had only emphasised the gradual *erosion* of continents, which was essential to the wise designfulness of the system he was advocating; and he left the nature of continental *elevation* so unclear that one of his followers, John Playfair, could argue for its gradual character while another, Sir James Hall, felt that sudden elevations were equally consistent with Hutton's basic intentions[42].) Cuvier believed he had good evidence that the most recent revolution had been a sudden incursion by the sea, of prolonged duration, followed by the re-emergence of the present continents. Because it was relatively recent, it had left the clearest traces and was easiest to interpret, and therefore could properly be used as a paradigm example of such events. Since the junctions between marine and freshwater strata in the Paris 'basin' were fairly abrupt, it was only natural for Cuvier to conclude that they bore witness to similar sudden changes at an earlier period. Likewise, since only a sudden change of environment could, he believed, have been responsible for the extinction of the most recent vanished fauna, so also a similar sudden event must have caused the extinction of the earlier fauna preserved in the freshwater strata of Montmartre. Thus the discovery of an alternation of marine and freshwater formations seemed to be striking evidence for the view that Earth-history had been punctuated by sudden changes in physical geography, each marine incursion being adequate to account for the extinction of a terrestrial fauna.

This conclusion was not only justified methodologically, since it used the more recent example to clarify the earlier and more obscure, but it also had the great merit of bringing these revolutions within the 'Newtonian' system of unchanging natural laws. However obscure the cause of revolutions might be, their repetition implied that they formed part of the ordinary course of nature. It is important to note that Cuvier was not forced into this concept of repeated revolutions by having to compress geological events into an inadequate time-scale. As we have seen, he was prepared to estimate the age of the Montmartre fossils in millions of years; and in his joint memoir with Brongniart he particularly emphasised that the uniform orderly sequence of strata and their large total thickness clearly implied extremely long periods of tranquil deposition. It was only at very infrequent intervals that these gradual processes, essentially similar to those occurring in present seas and lakes, had been interrupted by sudden changes.

Furthermore, Cuvier did not envisage these sudden changes as having been worldwide in their effect: on the contrary, he believed they had only affected one continental area at a time. This was suggested to him partly by the fact that the Montmartre fauna seemed to have closest zoological affinities with the present faunas of America and Australia—insofar as it resembled any living faunas at all. This provided him with a hypothesis for the re-population of each newly emergent land-mass, by migration from another continent that had not been affected by that particular revolution.

VIII

All these ideas were integrated and further developed in the *Preliminary Discourse* that Cuvier prefixed to his great *Researches on the fossil bones of quadrupeds* (1812)[43]. Three of the four volumes of this work consisted of new versions of the monographs that he had published during the previous years in the *Annals of the Museum*, on the osteology of various fossil species and of related living species of tetrapod vertebrates. The first volume contained, in addition to the *Discourse*, a greatly enlarged version of his joint stratigraphical memoir with Brongniart, accompanied by fine sections and a geological map; and the whole work was dedicated, significantly, to the astronomer Laplace, whose Newtonian *Celestial Mechanics* Cuvier clearly hoped his own work would emulate in the realm of terrestrial history.

The attractively written geological essay with which the work began was recognised at once as a work of the highest importance. It was later issued separately as a *Discourse on the revolutions of the surface of the globe*, which went through several editions and was translated into the other main European languages. In this way Cuvier's geological theory, which many years later was given the misleading name of 'catastrophism', became widely known and influential among the general reading public as well as among men of science. Cuvier himself rarely used the word 'catastrophes', for its overtones of disaster were largely extraneous to his conception of these regular and natural events; he preferred the term 'revolutions', with its more Newtonian flavour. Likewise, although the editor of the English editions, the Scottish geologist Robert Jameson (1774–1854), entitled the work *Theory of the Earth*, Cuvier himself always avoided this phrase on account of its associations with the earlier speculative systems he so much deplored.

In substance the *Preliminary Discourse* brought together the ideas that Cuvier had already suggested, and welded them into a coherent theory. Within an immensely lengthy time-scale of Earth-history, generally tranquil conditions similar to those observable at the present day had been interrupted occasionally by sudden major changes in physical geography. Much of the essay was devoted to showing that these revolutions must have been sudden, and that no observable process was adequate to account for them. This was not a reversal of the method of actualism, but simply an acknowledgement that "unfortunately . . . none of the agents that [Nature] employs today would have sufficed to produce its ancient effects". Certainly it did not imply that the cause of revolutions was unknowable, still less supernatural. Cuvier doubted whether the cause would be at all easy to discover, but he suggested that in the meantime "the most important geological problem" was to determine the exact character of these events, so that at least the nature of their cause could be more narrowly circumscribed.

Given such sudden events, the extinction of functionally well-adapted animal species was satisfactorily explained. The problem of the origin of each new fauna was interpreted in terms of intercontinental migration. Cuvier specifically denied that it was necessary to postulate the creation of new species, and he used the analogy of the present marsupial fauna of Australia to emphasise just this point. If a future revolution were to submerge Australia and destroy its fauna, and if it were later to re-emerge and be re-populated by the migration of placental mammals from Asia, a future generation of naturalists would *not* be correct in inferring that the newer fauna had been specially created to replace the extinct marsupials. In a way, of course, this explanatory device merely pushed the problem of the origin of species further back in geological time; but to Cuvier the important point was that this was a matter on which science had no evidence to offer and therefore no right to speculate. He denied, on the grounds already described, that Lamarck's hypothesis of evolution had any validity: above all, if it were correct "some traces of these gradual modifications ought to be found", and they were not; and furthermore the mummified animals from Egypt provided an actualistic test which proved that no perceptible modification had taken place in the last few millenia. Lamarck had only been prepared to concede the reality of extinction as an exceptional occurrence due to the recent influence of Man; and so Cuvier was particularly concerned to show that the rise of civilisation post-dated even the most recent revolution. Since there was a striking absence of human remains in the débris of that event, any men that might have been in existence previously must have been too localised and primitive to be responsible for the extinction of a whole fauna of large mammals. Such massive extinctions were thus established, in Cuvier's view, as events that were a regular and *natural* feature of the history of the Earth.

IX

When Robert Jameson edited the English translation of Cuvier's work[44], he changed not only the title but also its contents. He added lengthy editorial notes which tried to point out how Cuvier's most recent revolution could be identified as the scriptural Flood, thus confirming the historicity of that event from scientific evidence of the highest respectability. Since most British men of science (and most

English-speaking historians of science ever since) learnt of the theory of revolutions through Jameson's editions, it is hardly surprising that Cuvier was assumed to be supporting arguments that were in fact of far greater concern in Britain than they ever were on the Continent. Cuvier was a prominent lay member of the French Protestant community, and there is no reason to doubt the sincerity of his personal religious beliefs. But he was also a child of the Enlightenment, and he considered that science and religion should not interfere in each other's affairs, but should, for the good of both, be kept apart. Any attempt to support the truths of religion from the findings of science was, in his view, a futile and misguided enterprise. His theories thus owed nothing, at least on the conscious level, to a desire to make his science conform to scripture.

It is true that Cuvier mentioned the Pentateuch in the *Preliminary Discourse*, and that he saw no good reason why its authorship should not be attributed to Moses himself. However he evidently accepted the findings of the new biblical criticism, since he cited the German orientalist Eichhorn for the date by which the Pentateuch had reached "its present form". More significant is the fact that of some fifteen pages devoted to ancient records his discussion of the Old Testament occupied only one. In fact he used the Pentateuch merely as one of many ancient literatures that suggested the low antiquity of civilisation. Moses, "the legislator of the Jews", was merely cited because he was likely to have known, and used, any Egyptian records of a higher antiquity, if such traditions had then been current. Throughout this section of his argument, Cuvier's aim was to show that extinctions could *not* be attributed to Man's activities, as Lamarck supposed, but that they were part of the regular course of nature; and this involved emphasising the recent rise of civilisation. This also explains why he was so concerned to stress the *absence* of human fossils in the débris of the last revolution, and indeed the absence of any evidence for the co-existence of Man and the extinct fauna: for this helped to show that before the last revolution Man had been too primitive and localised to have been responsible for the extinctions. Had it been his intention to bolster a belief in a literally interpreted scriptural Flood, he would have been far more concerned to explain away the absence of human fossils.

However, when imported by Jameson into Britain, Cuvier's work was welcomed enthusiastically by those who were eager to find support for the authority of religion—and hence also support for the social order—from the authority of science. The progress of physical science had provided substantial 'arguments from design' for the natural theology of the eighteenth century; but the corresponding devaluation of evidential theology had, it seemed, only encouraged the deism and insidious scepticism that had led to the excesses of the Revolutionary era in France. What was needed, in order to restore the authority of revealed religion, was some independent scientific confirmation of the reliability of scripture. Since the new science of geology, like Christianity itself, was essentially *historical*, it was clearly in a special position—either to support or to undermine. These were the motives that lay behind the development of a new form of diluvial theory in Britain, based ostensibly on Cuvier's theory of revolutions. Cuvier's most recent revolution was thus welcomed as being none other than the scriptural Deluge. It was a *new* form of diluvial theory, in that it was based on a much wider range of scientific evidence than De Luc's had been. But it resembled De Luc's in that it was applied—unlike Woodward's still earlier theory—only to the superficial deposits, not to the regular strata; and it could be combined with an indefinitely extended time-scale for the pre-diluvial world, and indeed was so combined by almost all men of science. The shade of Ussher was still active among popular expositors of science and religion in England, but among those who had any pretensions as men of science the barrier of Ussher's time-scale had long been broken, at least for the pre-human epochs of Earth-history.

The transformation of Cuvier's localised last revolution into a unique universal deluge was carried out with greatest scientific authority by William Buckland (1784–1856), Reader in Geology at Oxford, whose popular and entertaining lectures gave the diluvial theory immense influence among English geologists. In his inaugural lecture in 1819, Buckland felt it necessary to defend geology against suggestions that it undermined the credibility of Christian revelation, and he argued that it actually supported religion by giving scientific evidence for the recent origin of man and for a universal deluge[46]. However, although he relied heavily on the scientific authority of "one of the most enlightened Philosophers, and the greatest Anatomist of this or any other age", his diluvial interpretation actually involved drastic *scientific* modifications of Cuvier's theory, for he had to show that the deluge had been transitory and universal, not prolonged and localised.

In effect, Buckland was following the intentions of De Luc and acting from the same motives; but he was trying to show that the scientific evidence witnessed to a deluge that was even closer to the literal scriptural account than De Luc's had been.

It is only fair to add, however, that soon afterwards Buckland believed he had found good scientific evidence for his interpretation. In 1821 a cave was discovered in Yorkshire with a rich deposit of fragmentary bones. Buckland used Cuvierian comparative anatomy to prove that the animals represented were extinct species; and then showed—after observing the habits of a hyaena in the Exeter zoo—that the bones had been gnawed by hyaenas and that the cave had been a hyaena den. From this he inferred, reasonably enough, that the area had been land *before* whatever event annihilated the extinct species, as well as since; and this implied that the most recent revolution had *not* been a general interchange in the positions of continents and oceans, as Cuvier (following De Luc) had suggested, but a more transitory event that had left the present continents where they were, with only superficial changes (Fig. 3.8). Buckland was right to cite the Huttonian Sir James Hall in support of this interpretation of the 'diluvial' deposits as the débris of some huge tidal wave[47].

Thus far Buckland's work was widely admired, and it won him the Royal Society's Copley Medal. But when in his *Relics of the Deluge* (1823) he went on to argue for the global universality of this event he met with far more criticism, for the diluvial deposits seemed to most observers to be confined to northerly latitudes (as indeed they are, in modern terms, as glacial deposits), and there was no evidence that they were of the same age in all areas[48]. When, finally, Buckland used reports of bones being found at high altitudes in the Andes and the Himalaya as evidence that the diluvial event had not been confined to low ground, as Cuvier believed, but deep enough to cover the highest mountains, there was a widespread and justified feeling that science had been twisted to conform to scripture—and that scripture had been twisted at the same time to conform to what Buckland regarded as science. This feeling was nicely epitomised by John Fleming (1785–1857), a Scottish naturalist and Presbyterian minister, when, in the course of a protracted argument in the scientific periodicals, he wrote a paper terming "The geological Deluge, as interpreted by Baron Cuvier and Professor Buckland, inconsistent with the testimony of Moses *and* the Phenomena of Nature" (my italics)[49].

Fig. 3.8. One of Buckland's illustrations (1823)[48] of a bone-bearing cave, showing the bones scattered in a deposit sealed in with a layer of stalagmite (C). Buckland argued that many such caves had been the dens of an extinct species of hyaena, and that a transient 'diluvial' tidal wave had caused the extinction of both the hyaena and of the other species on which it had preyed.

Fleming's coupling of Cuvier and Buckland under the same condemnation might seem to invalidate the distinction that has just been drawn between them. In fact, this was due to his use of one of Jameson's English editions of Cuvier's work, which, as we have seen, interpreted it in editorial notes in far more diluvial terms than Cuvier would have wished. On the religious aspect, Cuvier would probably have agreed with Fleming that Buckland's "indiscreet union of Geology and Revelation" deserved Francis Bacon's censure as "*Philosophia phantastica, Religio haeretica*", and that true religion could afford to reject "such faithless auxiliaries" as Buckland's science seemed to Fleming to be. It was 'faithless' because it was patently untrue, in Fleming's opinion, even in purely scientific terms. Buckland's conclusions simply did not follow even from his own evidence, still less from other facts that Fleming felt he should have mentioned.

Fleming's criticisms of Buckland's deluge were indeed so radical that they tended to undermine even the Cuvierian concept of revolutions as well; but Fleming was untypical in carrying his attack so far, and to most men of science the main substance of Buckland's work seemed difficult to fault. It certainly suggested that the last revolution, and hence probably the earlier ones too, had been due to some kind of violent tidal wave sweeping briefly over the continents, annihilating the previous terrestrial fauna, excavating the valleys, and depositing the strangely unsorted superficial deposits of boulder-clay as well as the valley gravels. With the benefit of later research we can see how natural it was to interpret all the manifestations of the glacial period in these terms, for they were indeed causally unrelated to the present landscape, and there was little reason to suspect that a glacial agency might have been responsible. Buckland's concept of the last revolution as a transitory tidal wave could be, and in fact was, accepted widely, without agreeing with his much more questionable arguments for its universality. Thus Cuvier himself seems to have welcomed it, without abandoning his own view that revolutions were only localised events. He had urged geologists to try to define the *nature* of revolutions more closely, in order to circumscribe more closely their possible *cause;* and this was exactly what Buckland had tried to do. Buckland did not attempt to suggest a cause for his deluge. However, like Cuvier, this was not because he thought the cause was unknowable: on the contrary, he said that he and other writers had "not *yet* shown by what physical cause it was produced" (my italics); and he later planned a sequel to the *Relics*, which was to have been devoted to a discussion of the physical cause (the second volume was never written, because he became convinced that the cause had been glacial and not aqueous: there had indeed been a revolution, but it had not been a deluge).

X

Another promising way of defining the nature of revolutions more precisely, and thereby circumscribing their cause more closely, was to fix their occurrence within the geological timescale. As Cuvier pointed out, one of the most urgent needs in geology was for more accurate and reliable stratigraphy. Quite apart from the spur of his scientific authority and the example of his own and Brongniart's work, the rapid development of stratigraphical studies was greatly aided by the final restoration of peace throughout Europe in 1815. Scientific travel was once more possible without political restrictions, and contacts between men of science in different countries flourished as never before. International comparisons of strata could thus be made with much greater reliability. At the same time this work was facilitated by the growing use of geological maps. Although geological maps of various kinds had been used extensively in France for many years (Lavoisier, for example, had taken part in the state survey of mineral resources before the Revolution), Cuvier and Brongniart's map of the Paris region (1811) became the pattern for all local stratigraphical memoirs in subsequent years: Freiesleben, for example, included a fine map in the last volume (1815) of his stratigraphy of Thuringia. In the same year William Smith was at last able to publish his huge map of England and Wales, covering a larger area than any map published previously; and his subsequent publication of drawings of the fossils characteristic of each formation helped to draw attention to the empirical value of fossils in correlation (Fig. 3.9)[50]. The comparison of stratigraphical successions in different countries tended to emphasise the recurrence of even the most distinctive rock-types, so that this earlier criterion for correlation came to be supplemented more and more by the use of fossils.

At the same time the other earlier criterion, that of topographical position, proved to be irreconcilable with the use of fossils. Brongniart, for example, showed in a memoir *On the zoological characters of*

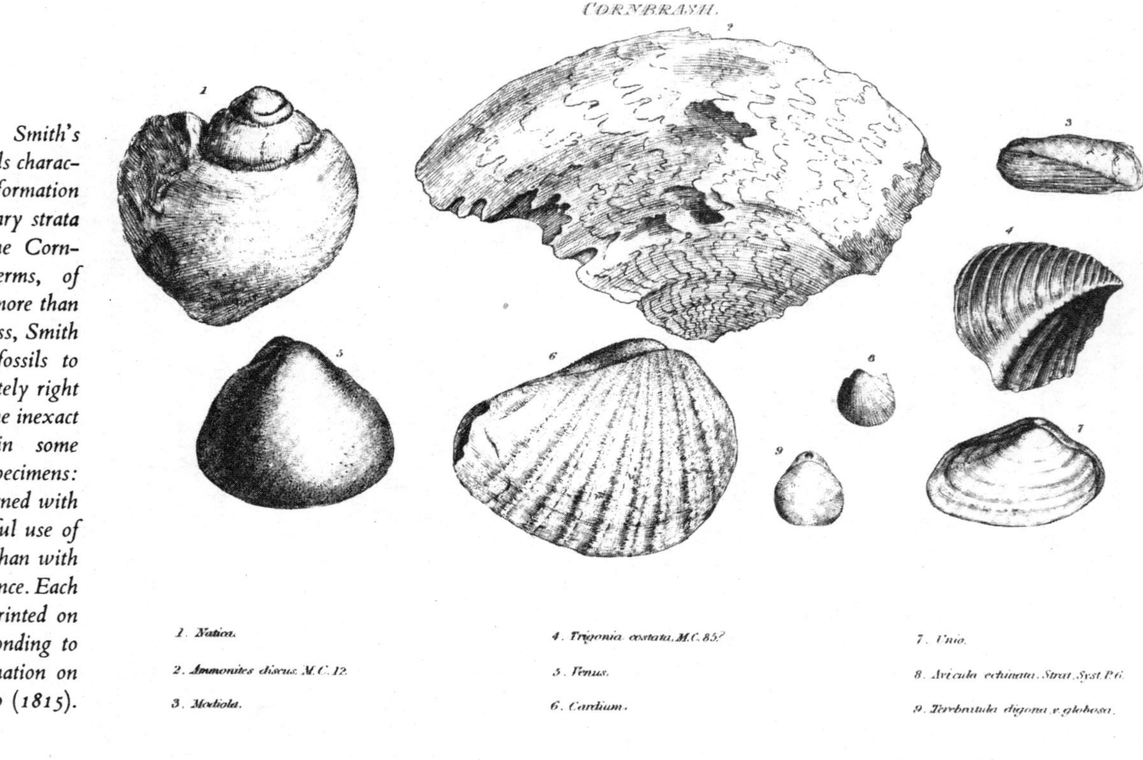

Fig. 3.9. One of Smith's illustrations of the fossils characteristic of a particular formation in the English Secondary strata (1819)[50]. Although the Cornbrash (in modern terms, of Jurassic age) is rarely more than a few metres in thickness, Smith was able to use its fossils to trace its outcrop accurately right across England. Note the inexact identifications and (in some cases) fragmentary specimens: Smith was more concerned with the empirically successful use of fossils in stratigraphy than with their biological significance. Each set of drawings was printed on coloured paper corresponding to the colour of that formation on Smith's geological map (1815).

formations (1821) that the distinctive fossils of the Chalk could be found in a hard black limestone outcropping more than 2000 metres above sea-level in the Savoy Alps; and in a second memoir two years later he showed that the Tertiary fauna of the Paris region could likewise be found at high altitudes in the Vicentine Alps[51]. These results were of major theoretical importance, not only because they implied that fossils, rather than lithology or altitude, should be the primary criterion for correlation, but also because is became clear that mountains did not all date from the earliest periods of Earth-history. Even high mountains like the Alps had evidently been elevated at a comparatively recent epoch, that is, subsequent to the period of deposition of the Tertiary strata around Paris.

It is hardly surprising that these remarkable results should have been integrated into the Cuvierian theory of occasional sudden revolutions. The increasing emphasis on fossils in stratigraphy led to the recognition of major faunal discontinuities between many of the main formations—for example between the Chalk and the Tertiary strata. It thus seemed that marine faunas, as well as the terrestrial vertebrate faunas that Cuvier had reconstructed, had been affected by revolutions of some kind. Since faunal discontinuities seemed to punctuate the fossil record at intervals as far back as it had been deciphered, this appeared to confirm Cuvier's concept of occasional revolutions as events that were natural to the Earth's history. Furthermore, Brongniart's demonstration of the relatively recent elevation of the Alps suggested a possible mechanism for these revolutions. It seemed hardly conceivable that mountain elevation, with the astonishing contortions of strata that it manifestly involved, could have occurred by any gradual process. If, however, it occurred abruptly, by the sudden release of internal stresses in the Earth's crust, it could well have profound effects on the environment of organisms at the surface. Even a geologically small seismic disturbance, such as the famous Lisbon earthquake of 1755, could produce a tidal wave with catastrophic effects in distant areas; so on this actualistic analogy the far larger disturbance caused by the elevation of a mountain range could have had drastic effects on an almost world-wide scale.

This was the theory that was developed by one of Cuvier's most brilliant geological disciples, Léonce Élie de Beaumont (1798–1874). From a detailed study of areas of folded rocks throughout Europe, he proved that mountain-building episodes had occurred at many different periods of Earth-history. Each episode could be dated, with the aid of the latest stratigraphical research, by noting the ages of the youngest strata involved in the folding and of the oldest strata overlying them in an undisturbed state: between these two periods the mountains must have been elevated. But it then seemed that these episodes of mountain-building coincided with the major faunal discontinuities between formations: for example the Pyrenees appeared to have been elevated between the Chalk period and the early Tertiary. This implied that the sudden elevation of mountain ranges could have been the immediate cause of the revolutions that had produced such massive extinctions of both marine and terrestrial organisms. As for the cause of the elevations themselves, Elie de Beaumont believed that the sudden release of internal stresses was a likely explanation, not least because the mountain ranges elevated at any given date appeared to have a characteristic trend, as might be expected on mechanical grounds if the crust had buckled under stresses acting in a particular direction. Élie de Beaumont's *Researches on some of the revolutions of the globe* (1829–30)[52] gained an enthusiastic reception, for they seemed to synthesise several hitherto unrelated problems into a coherent theory, and to bring Cuvier's hope of discovering the physical cause of revolutions nearer to fulfilment. Although the over-geometrical treatment of the directions of mountain systems was quickly criticised, the theory did appear to offer a satisfactory explanation of the apparently discontinuous nature of the history of life.

XI

Meanwhile, however, the development of stratigraphy, and of Cuvierian palaeontology, had been shaping a new concept of the history of life as not only punctuated by occasional discontinuities but also in some sense *progressive*. As we have seen, Cuvier was averse to the idea of a single scale of beings on philosophical grounds, but he also believed it could be disproved empirically by anatomical comparison. The animal kingdom was, he maintained, divisible into four major 'branches' (*embranchements*): within each branch anatomical comparison was possible and indeed fruitful, but between them (for example between vertebrates and molluscs) there was no valid comparison. This precluded the arrangement of all animals in a single serial order;

but the animals *within* a branch could properly be placed in an order that reflected some kind of genuine 'scale'. Thus within the vertebrates Cuvier arranged four classes from mammals down to fish, and within the molluscs four classes from cephalopods down to 'acephala' (that is roughly, bivalves)[53]. It was this limited kind of scale that seemed to be reflected, at least for the vertebrates, in the sequence of the fossil record.

The majority of the vertebrates reconstructed in Cuvier's *Researches on fossil bones* were the mammals from river-gravels and other superficial deposits. Most of these were of known genera but unknown species (for example the mammoth). Next in abundance were those from the Tertiary strata at Montmartre, mostly mammals but with a few reptiles and fish, and even a bird; most of the genera here (for example *Palaeotherium*) were totally unknown among living animals. However, apart from the bones of a few marine mammals—seals and sea-cows—in the underlying *Calcaire grossier*, no mammalian remains whatever had been found in any earlier strata. Among the Secondary formations Cuvier had shown that Faujas's giant crocodile from the Chalk was in fact a hugh lizard-like marine 'saurian' (later termed a mosasaur); he had proved that a bird-like fossil from Bavaria was a flying reptile ('*ptero-dactyle*'); and he knew of another lizard-like reptile from the much older (Permian) strata of Thuringia. In the *Preliminary Discourse* he therefore felt confident in stating, as one of the 'laws' that his research had defined, that "it is certain that the oviparous quadrupeds appeared much earlier than the viviparous". 'Lower' vertebrates had appeared before those 'higher' in the scale; and though Cuvier himself avoided such terms on account of their associations with a broader concept of a scale of beings, he had implicitly recognised and sanctioned the interpretation of the fossil record in these 'progressive' or at least directional terms.

It was an interpretation that received cumulative support in the following years. The secondary fossils that Cuvier had tentatively identified as the remains of extinct crocodiles proved to be another new fish-like marine reptile, later termed an ichthyosaur. The English parson-geologist William Conybeare (1787-1856), emulating Cuvier's methods, reconstructed yet another strange marine reptile (the plesiosaur) from the English Secondary rocks (Fig. 3.10). Another amateur English geologist, the physician Gideon Mantell, soon afterwards discovered in Sussex some fossil teeth, which Cuvier suggested had

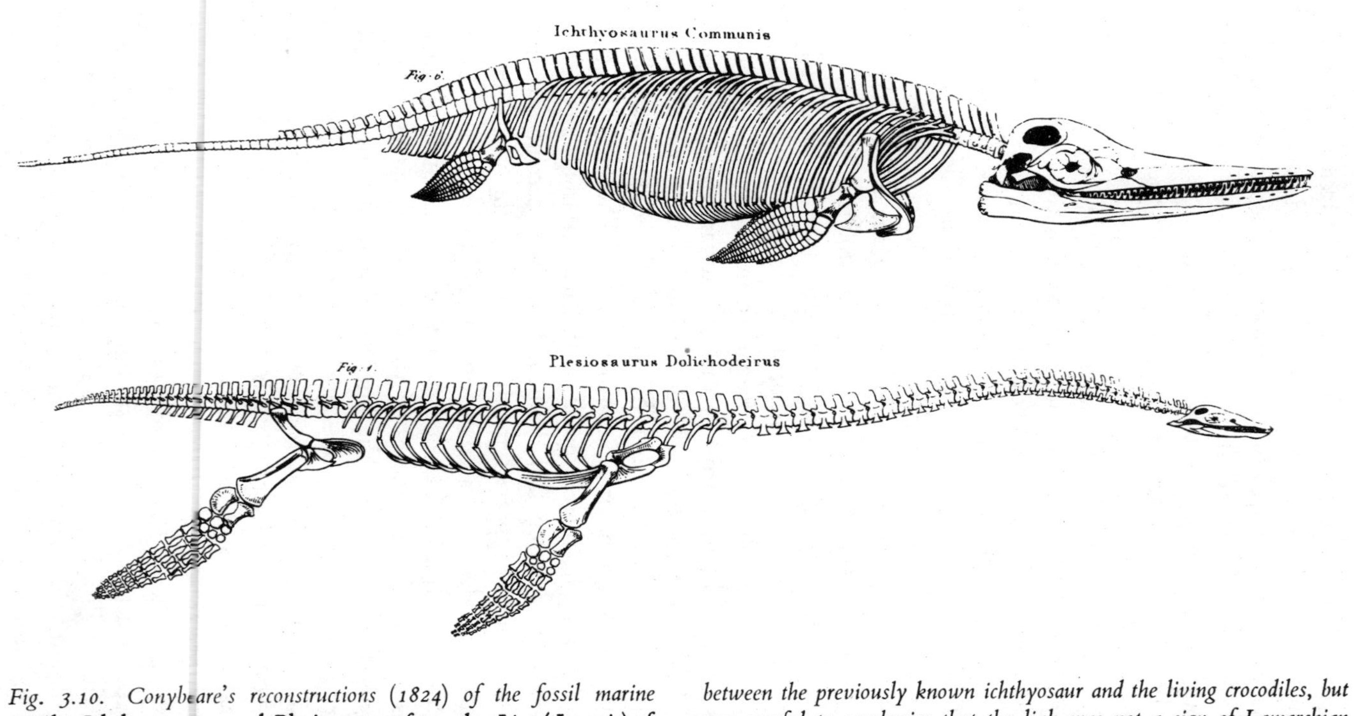

Fig. 3.10. Conybeare's reconstructions (1824) of the fossil marine reptiles Ichthyosaurus *and* Plesiosaurus *from the Lias (Jurassic) of England*[54]*. He welcomed the discovery of the plesiosaur as a 'link' between the previously known ichthyosaur and the living crocodiles, but was careful to emphasise that the link was not a sign of Lamarckian transmutation.*

belonged to an unknown herbivorous reptile; and this later proved to be the first *terrestrial* reptile from Secondary strata, the huge dinosaur *Iguanodon*. From such discoveries there emerged a concept of the Secondary era as an age of abundant and ecologically diversified reptile faunas, in which mammals had been conspicuous by their absence[54].

Even an apparent anomaly to this rule turned out to give it unanticipated support. One of Buckland's pupils gave him a specimen of an apparently mammalian jaw, which had been found in Secondary strata at Stonesfield in Oxfordshire. While on a visit to England Cuvier confirmed that it was indeed mammalian, but that its affinities lay closer to the opossum than to any placental mammal. The discovery of any mammal whatever in Secondary strata was so unexpected that attempts were made to throw doubt on both its stratigraphical position and its zoological affinities[55]. In time, however, it was recognised as being not an anomaly at all, but a confirmation and refinement of the parallelism between the fossil record and the 'scale' of vertebrate classes: for the Stonesfield marsupial had appeared, as expected on that model, *after* the reptiles but *before* the more 'advanced' placental mammals (modern research has shown that these Jurassic mammals were even more primitive than the marsupials, but this would not have affected the conclusion).

This refinement of the parallelism to categories *within* a single class seemed to be confirmed by the example of man himself. Being obviously of the greatest intrinsic interest, the most thorough search was made for any fossil remains of human beings; but in one case after another alleged human fossils proved on anatomical study not to be human at all, or not to be genuine fossils. For example, Cuvier found that Scheuchzer's famous "man a witness of the deluge" was nothing but a large amphibian; and the only human fossil he could cite was an obviously recent skeleton from lithified coral sand on Guadaloupe. Man as the 'highest' of the mammals thus seemed to date from a far more recent period than the other placental mammals (evidence that early Man had co-existed with the extinct 'diluvial' fauna was generally discounted)[56].

This growing sense of a progressive *direction* to the history of life was then confirmed unexpectedly from studies of fossil plants. The technical problems of fragmentary material were even greater here than they had been for vertebrates, for the reproductive structures on which the classification of plants was based were very rarely preserved in the fossil state, and the fragmentary leaves and stems that were preserved were extremely difficult to identify. However the German naturalist Ernst von Schlotheim (1764–1832), who had studied biology with Blumenbach and geology with Werner, made a pioneer attempt to apply Cuvierian comparative methods to palaeobotany, reconstructing the fossil flora of some Secondary (in modern times Carboniferous) strata in Thuringia. In his *Contribution to the flora of the former world* (1804)[57] he recognised that although his fossil plants had a general resemblance to the tree-ferns of living tropical floras, in detail they were of unknown kinds, and he concluded that they represented a totally extinct flora—a conclusion that was closely parallel to Cuvier's at about the same date.

This result was greatly amplified later, after the spectacular development of stratigraphy in the first two decades of the century, in the work of Alexandre Brongniart's son, Adolphe (1801–1876). His *Forerunner of a history of fossil plants* (1828)—the very fine monograph to which this was introductory was never completed—summarised the preliminary results of his research on fossil floras of many different geological dates[58]. He concluded that four distinct periods could be defined in the history of plant life. Between these periods there were abrupt floral discontinuities, but within each the changes were more gradual: this was similar to the conclusion that had been reached by this time for the history of animal life. Furthermore, like the 'progress' of animals in the course of time, so also among plants the successive periods were marked by increasing diversity and increasing complexity in the groups represented. Thus the first (in modern terms, Upper Palaeozoic) period was dominated by vascular cryptogams; in the second, with only a poor flora, there were the first conifers; in the third (roughly, Mesozoic), the first cycads had appeared and, with conifers, comprised about half the total flora; and finally, in the fourth (Cainozoic) period, dicotyledons had made their first appearance and had come to dominate the flora. These "positive results, independent of all hypothesis and all preconceived theory"—or so he asserted—were to Brongniart a gratifying confirmation of the 'progressive' model of the history of life that had been emerging from the study of animal fossils.

However the younger Brongniart suggested further implications of his work. The tropical character of the flora of the first period (that is, mainly of the Coal strata) could now, he believed, be regarded as firmly established. Although the genera were extinct, the profusion of giant tree-ferns, club-mosses and horse-tails could in general ecological character only be matched at the present day in the hot damp rain-forests of the tropics (Fig. 3.11). Brongniart concluded from this actualistic analogy that the climate of the Coal period had been as hot as, or even hotter than, the present tropics; and the later floras seemed to him to confirm that there had been a gradual reduction in the temperature of the globe.

This conclusion was an exciting confirmation of a contemporary geophysical theory, which maintained that the 'progress' of life had been paralleled by a directional development of the inorganic environment. A knowledge of the geothermal gradient observed in mines had been part of the empirical basis for Buffon's theory of the gradual cooling of the globe; but the validity of both the observations and their interpretation had been criticised, and ideas of a hot origin of the Earth had almost dropped out of scientific discussion. Then, in the 1820s, more detailed studies showed that the gradient was genuine and indeed universal; and when the mathematical physicist Fourier applied his equations of heat flow to the problem of a cooling sphere, and concluded that the geothermal gradient was due to the Earth's *residual* heat, Buffon's model of Earth-history was immediately restored to scientific respectability. Not only did it sanction, with all the prestige of physics, any geological evidence for a gradually cooling surface temperature, but it also implied that the *rate* of cooling had itself diminished gradually[59]. Thus for example it could explain why the scale of mountain-building and of volcanic activity seemed to have been much greater in the early periods of Earth-history than in more recent epochs, a diminution of intensity that made actualistic comparisons with the present of limited value in the study of the more ancient revolutions of the globe.

Applied to the history of life, this theory suggested an ecological or even physiological explanation of the progressive increase in the diversity and complexity of the organisms that had existed at successive periods. The simpler or 'lower' animals and plants of early periods had been adapted to the hot conditions then prevalent; later, as the mean temperature of the Earth's surface dropped and more temperate climates had become available, 'higher' animals had been able to exist; while the increasing range of available habitats had been matched by

Fig. 4.11. Adolphe Brongniart's illustrations (1828–37) of parts of the fronds of a fossil tree-fern from the Coal Measures, and of a modern tropical tree-fern for comparison[58]. **Brongniart's careful description and ecological reconstruction** of the Carboniferous flora suggested that the Earth's surface had been generally hotter at that period in Earth-history: this was in accordance with the contemporary geophysical theory of a gradually cooling Earth.

the increasing diversity of the fauna and flora. Brongniart also suggested a physiological correlation between the luxuriance of the Coal forests and the later appearance of terrestrial vertebrates. The size and profusion of the plants suggested a high concentration of carbon dioxide in the atmosphere at the Coal period: only when this had been reduced by the locking-up of carbon in the coal deposits would the atmosphere have become suitable for the air-breathing reptiles of the subsequent period, while a still higher proportion of oxygen would have been required before the more active mammalian organisation could exist. Whatever the merits of Brongniart's arguments (some of them have a surprising similarity to modern theories of the evolution of the atmosphere), his theory exemplifies the concern of scientists in the 1820s to integrate the results of geological and palaeontological research. Both fields seemed to converge towards a concept of the history of the Earth and of life that was above all *directional* in character[60].

XII

There was, however, one major problem that this synthesis failed to explain. As we have seen, Cuvier's theory of revolutions was primarily designed to account for the apparent fact of mass extinctions, and in the earlier years of his research it had been reasonable for him to bypass the problem of the *origins* of new faunas, by attributing their appearance to migration from other continents. But as stratigraphy developed with increasing emphasis on the fossil contents of strata, so it became apparent that the sequence of fossil remains was broadly similar in all parts of the world. The question of the origin of new faunas and new species, which Cuvier had at first legitimately considered to be outside the proper realm of scientific questions, was thus pushed inexorably into the domains of 'positive science'. Ironically it was the very success of Cuvier's methods of reconstructing ancient faunas that changed the status of the question, because the growing evidence for a 'progressive' succession of life implied that the origins of new forms must have occurred at intervals during geological time. The origins of species and, far more seriously, of the major types of organisation embodied in the classes of the animal and plant kingdoms, were no longer veiled in the impenetrable obscurity of the primordial origins of the Earth itself: they were brought forward in time to definable points in the record of the strata. The fish and reptilian types of organisation appeared to date only from the older part of the Secondary era (though the stratigraphy here was still somewhat obscure); more positively the mammalian organisation had apparently come into existence only at the beginning of the Tertiary epoch, or, allowing for Stonesfield marsupials, in late Secondary time at the earliest. Cuvier himself acknowledged these facts, as we have seen; but although he greatly enlarged the scope of his *Researches* in the years after its first edition was published, he continued in effect to ignore the problem, and to regard it as unscientific, until shortly before his death he was goaded into an outspoken public denunciation of a solution that seemed to him to imperil the very foundations of his work.

Although Lamarck's concept of continual and 'progressive' organic flux had been formulated within an older natural philosophy, there was one member of the Museum staff, of Cuvier's own generation, who had been impressed by it. In their early years at the Museum Geoffroy St. Hilaire had been a close personal friend and scientific collaborator of Cuvier's, but they had gradually drawn apart as fundamental differences of outlook had become apparent between them—accentuated by the scientific arrogance that success bred in Cuvier, and by the speculative inclinations that Geoffroy acquired from the 'romantic' Nature-Philosophy (*Naturphilosophie*) of German naturalists. In their zoological work both had been concerned primarily with anatomical comparison; but whereas Cuvier's approach was basically functional, Geoffroy was more interested in formal (or, in later terminology, homological) comparisons[61]. Where Cuvier formulated a principle of the functional 'correlation of parts', Geoffroy defined a principle of homological 'connections' between organs, such that their relative positions always remained the same whatever their degree of development. The diversity of animals was to Cuvier primarily a manifestation of their varied functional and ecological adaptations; to Geoffroy it displayed the varied topological transformations in which a single anatomical 'ground-plan' could exist. However, these two divergent attitudes to anatomy—which were later to be fused within Darwin's theory—were at first applied only to the static problem of animal form, not to the question of the temporal origin of the diversity. It was Geoffroy who had collected the mummified animals in Egypt, and he seems to have been convinced at first that they were indeed a conclusive disproof of Lamarck's theory. It was only later, when his main interests had shifted on to problems of embryonic development, that he became convinced that animal organisation was more labile than Cuvier's principles allowed.

It had long been recognised that 'higher' vertebrates passed during embryonic development through stages that were reminiscent of 'lower' classes, and this had become an important argument for the reality of the scale of beings. But the 'preformationist' theory of development which this supported—with its concept of the initial 'germ' of each individual as being (in modern jargon) 'programmed' deterministically to ascend the Scale during development—was difficult to reconcile with the observation that occasionally development resulted in 'monstrosities'. This suggested to Geoffroy that only an 'epigenetic' theory could bring these phenomena within the scope of natural law: evidently external influences could somehow affect development, which therefore could not be rigidly programmed from the start. In his *Anatomical Philosophy* (1818–22)[62]—the title suggests a desire to emulate Lamarck's work—Geoffroy combined a study of the homologies of the respiratory organs with a study of human teratology; and it was these two areas of research that he then sought to fuse into a single concept of animal form as indefinitely labile. In a series of experiments at a commercial chick-hatching establishment near Paris, he found that monstrosities could be produced at will by appropriate alterations in the external environmens of the eggs at appropriate stages in their incubation. This seemed to him to confirm the "general axiom" that "there is nothing fixed in nature", an assertion that he made explicitly with reference to Lamarck's work; and he went on to maintain that this general principle of flux was particularly applicable to organisms, "the essence of which lies in effect in the transformation and metamorphosis of their parts".

Such a Lamarckian definition of life had obvious implications for the *history* of life. These implications Geoffroy chose to make explicit by attacking Cuvier on his own ground, with a rival interpretation of the living and fossil crocodiles (1825). Was it not possible, the title of his memoir suggested, that the living species had "descended, by an uninterrupted path of generation", from the fossil species?[63] Using Cuvier's own methods, he argued that the species had been adapted to different environmental conditions; but he drew from this the most un-Cuvierian conclusion that these anatomical differences could have been induced by the changes in the environment itself. What his chick experiments had proved to happen "in miniature and under our eyes" could surely have occurred on a larger scale as a result of the more drastic changes postulated by physicists and geologists.

What Geoffroy was doing here was, in effect, to take his own concept of animal organisation as indefinitely labile under environmental influences, and to integrate it with the contemporary geological theory of a directionally changing environment. The directional model of Earth-history that was now available made it unnecessary for him to adopt Lamarck's notion of an inherent progressive tendency in life itself; apart from the general belief in nature as flux, all that Geoffroy had to borrow from Lamarck was his secondary mechanism of the modifying influence of the environment. Furthermore, Geoffroy could even use the geological theory of sudden revolutions—which Lamarck abhorred—as a means of explaining the abrupt transitions from one type of animal organisation to another: such episodes of drastic environmental change would, on the actualistic analogy of his experiments, produce correspondingly large changes in anatomy and physiology. It was thus an evolutionary theory that, unlike Lamarck's, provided for a variable *rate* of change, and which in principle was perfectly concordant with contemporary geological theory.

However, despite their integration with Cuvier's geology, Geoffroy's ideas were totally unacceptable to Cuvier on biological grounds, for they conflicted with his belief in the necessary functional stability of animal organisation as fundamentally as Lamarck's theory had done. Moreover, Cuvier can hardly have been encouraged to take Geoffroy's theory seriously when, in a later and more general exposition (1828), Geoffroy cited in its support a "progressive series" of fossils that showed no sense of anatomical affinities and no understanding of relative geological ages (the Chalk mosasaur was *followed* by the Oolite teleosaur, which led immediately to the megalonix of Superficial deposits placed *before* the early Tertiary palaeotherium!)[64]. Finally, when in 1830 Geoffroy publicly approved a paper suggesting anatomical homologies between fish and cephalopods, Cuvier felt that such irresponsible speculations had gone too far, and he launched an intemperate attack on Geoffroy, on the recently-deceased Lamarck, and on all manifestations of the 'unscientific' Nature-Philosophy. The suggestion that the anatomy of the 'lowest' class of vertebrates could be linked to that of the 'highest' class of molluscs was not only opposed to Cuvier's demonstration of four radically distinct branches in the animal kingdom; more fundamentally, its implication that all animals could be linked in a single serial Scale cut at the roots of his biological methods and philosophy, and in effect attacked the value and meaning of his whole life's work. The acrimonious argument between Geoffroy and Cuvier was only peripherally a debate between evolution and the fixity of species: fundamentally it was a clash between opposing philosophies of nature[65]. Nevertheless, precisely because Cuvier attacked Geoffroy's work as a threat to the good name of anatomical science, and attacked it moreover from his well-earned position of immense scientific authority and prestige, his victory in the argument served to brand evolutionary theories as speculative and unscientific for the next thirty years. Only against this background can Darwin's extreme caution in propounding his own theory be fully understood.

XIII

There was, however, a further reason why Cuvier's stand against evolutionary theories should have been so influential, especially in England. His concept of the functional integration and stability of the organism had historical roots not only in Aristotelian biology but also in the tradition of natural theology that stretched back to Ray and beyond. It was Ray who had used the astonishing 'designfulness' of organisms as his most persuasive example of *The wisdom of God manifested in the works of creation* (1691); and as the theology even of the orthodox had become more and more deistic in character in the course of the eighteenth century, such 'arguments from design' had become the popular commonplaces of natural theology. The intricate mechanisms of animals and plants all witnessed to the existence of a wise designer; but this argument necessarily placed a premium on the constancy of the organisation in which the design was embodied. Just as God held the physical world in constant orderliness through the 'secondary' agency of the natural laws of physics, so also organisms were held in constant designfulness by the laws of generation (that is, in modern terms, of heredity). This implied that the distinctive designs of different organisms must have been implanted at the creation, like the physical laws of matter. To enquire about the origin of species or of organic design was thus as far outside the realm of natural science as to enquire about, say, the *origin* of the laws of gravitation.

This attitude to organic design is well illustrated by William Paley's attractively written book on *Natural theology* (1802). This was a deservedly popular compendium of examples drawn from Ray and other earlier works; but although immensely influential in England throughout the early nineteenth century, scientifically it belongs to the eighteenth. The world it analysed was still one of static order; and the 'evolutionary' theory it opposed was the broadly Lamarckian speculation of Erasmus Darwin (Charles's grandfather). Even its opening page showed that Paley was totally ignorant of the new scientific evidence for the *history* of the Earth; for Paley assumed that it was meaningless to ask questions about the *origin* of a stone found lying on the ground, whereas if a watch were found beside it such questions would lead naturally to a consideration of the watch's designer. However when, in the following years, the new geological evidence was joined by a growing awareness of the history of life itself, this Paleyan tradition of natural theology became linked, almost inevitably, to the assumption that species had originated by some kind of 'creative' process acting *within* geological time. Having made such a large investment in the argument from design in organisms, natural theology was obliged to defend it against any scientific theory that seemed to derogate from the divine workmanship embodied in each and every species. To condemn the theories of Lamarck and Geoffroy was thus to condemn not only unscientific speculations but also materialistic ideas that were subversive of the religious framework of intellectual life.

How far, if at all, this motive lay behind Cuvier's belief in the stability of organisms it is impossible to say; for so firmly did he maintain the necessity of a rigid separation of science and religion that he never committed himself clearly on such matters. But when imported into the very different intellectual and religious climate of England, scientific work in the Cuvierian tradition acquired new overtones, not only, as we have seen, from the concerns of evidential theology, but also from the equally powerful concerns of natural theology. In effect, the older notion of the plenitude of nature was retained, but in a modified form that allowed it to be concordant with the results of scientific research. It was now clear that not all forms of life that *had* existed still *did* exist; but it could be assumed nevertheless that all forms that *could* exist had done so at some time or other. Plenitude was thus preserved *sub specie aeternitatis*. Furthermore, just as the older static concept of the scale of beings had been regarded as a manifestation of the divine wisdom in creating a hierarchy of all possible forms of life, so the new scientific evidence for the 'progress' of life (or at least of the vertebrates) in the course of geological time could now be regarded as a *temporal* actualisation of that divine plan.

Conybeare's memoir on the plesiosaur is a representative example of these concerns in action in the work of an English palaeontologist. Conybeare said that his newly discovered reptile, being "a link between the ichthyosaurus and the crocodile", formed "a transition between different races", and added to "the connected chain of organized beings"; yet he felt impelled at once to dissociate this notion of a Scale from its "absurd and extravagant application" to an evolutionary theory. But it is immediately apparent that his criticism of Lamarck was more than merely scientific, for he characterised the scale of beings as a "striking proof of the infinite riches of creative design", and asserted that even to consider a Lamarckian theory required "nothing less than the credulity of a material philosophy"[66].

It tended to be assumed, in earlier historiography, that such theological concerns had had a wholly baneful influence on scientific research. It is therefore important to point out that, like the metaphysical presuppositions underlying the science of every period (including our own), natural theology did not blind men to the 'correct' solutions of scientific problems, but exercised its influence rather in the choice of problems to be tackled and in the kinds of solution that were regarded as satisfactory. Thus the emphasis on the divine designfulness of each species certainly created a powerful (and partly unconscious) opposition to any evolutionary theory, and tended to discourage speculation about possible mechanisms for the origin of species. Yet at the same time it acted as an equally powerful incentive for the functional analysis and ecological reconstruction of individual fossil species (and of living species too), because it led naturalists to *expect* to find evidence of adaptive mechanisms in the construction of organisms—as the example of Conybeare's own work demonstrates. Similarly the interpretation of the temporal progress of life in terms of the temporal actualisation of a divine plan tended to discourage speculation about possible mechanisms for the emergence of 'higher' forms of life; yet at the same time it provided an equally powerful incentive to search for further evidence of organic progress, and it was the successful outcome of this search that gave Darwin's later evolutionary theory the empirical foundations that its acceptance required.

XIV

By about 1830, therefore, the spectular success of some three or four decades of research on fossils had transformed Cuvier's early demonstration of a single recent organic revolution into a palaeontological synthesis of very wide scope and explanatory power. The geological time-scale was firmly established as almost unimaginably lengthy by the standards of human history, yet documented by an immensely thick succession of slowly deposited strata. The successive formations of strata, and even in some cases individual strata, were clearly characterised by distinctive assemblages of fossil species, which enabled them to be identified and correlated over very wide areas. This correlation proved that in its broader outlines the history of life had been the same in all parts of the world.

Cuvier's recent 'revolution' had turned out to be only the last of many similar events punctuating the history of life, and having apparently sudden and drastic effects not only on the terrestrial faunas that he had reconstructed, but also on the more abundant marine faunas and on plant life as well. These revolutions were evidently natural events that were somehow built into the physical constitution of the globe, and in character they appeared to be sudden transitory tidal waves that swept occasionally over at least the low-lying areas of the continents. Their physical cause remained uncertain, but seemed to be connected with the occasional sudden elevation of mountain ranges, which in turn was interpreted in terms of the sudden release of crustal stresses. In any case, as drastic episodes of environmental change they were adequate to explain the mass extinction of faunas and floras of well-adapted species.

Superimposed on this picture of occasional sudden revolutions, a directional or 'progressive' element had emerged from the study of the history of life and of the Earth itself. There seemed to be a sense of progress behind the successive appearance of fish, amphibians, reptiles, marsupials, placental mammals and finally man himself; and this was paralleled among the plants by the succession from cryptogams to dicotyledons. This sense of progress in both animal and plant kingdoms seemed to be correlated with a directional development of the Earth itself. The great prestige of physics supported the view that the Earth had gradually cooled, from an incandescent origin through hypertropical and tropical phases towards its present relatively equable but climatically diversified state. The Earth's surface had thus presented a sequence of progressively more equable and more diversified environments, to be the habitats of progressively more complex and more diverse forms of life.

This general consensus of scientific opinion needed no explicit or large-scale formulation. The days of grandiose systems seemed to be over; and with the expansion of scientific activity and the proliferation of scientific periodicals this synthesis of biology and geology was expressed, as a similar consensus would be today, principally in the network of cross-references in a multitude of special monographs, memoirs, and 'letters to the editor'. It was a research tradition within which most of the results of this mass of detailed research could be fitted without difficulty. Only in two respects did it fail to provide adequate explanations of these results: for it left unclear the means by which individual species and—more important—the major types of animal and plant organisation, had come into being.

REFERENCES

1. G. Cuvier, 'Mémoire sur les espèces d'Elephans tant vivantes que fossiles', *Magasin encyclopédique*, 2me année, vol. 3, pp. 440–5 (1796). This preliminary publication gives the date of reading as 15 Germinal (3 April); the date 1 Pluviose is taken from the later full version in the Institute's own publication (*see note 12*).

2. See Henri Daudin, *Cuvier et Lamarck. Les Classes zoologiques et l'Idée de Série animale (1790–1830)*, Paris, 1926: chap. 1; Yves Laissus, 'Le Jardin du Roi', 'Les Cabinets d'Histoire naturelle', *in* René Taton (ed.), *Enseignement et Diffusion des Sciences en France au XVIIIe siècle*, Paris, 1964: pp. 286–341, 659–712; Joseph Fayet, *La Révolution française et la Science 1789–1795*, Paris, 1960: part 1, chap. 9.

3. See William Coleman, *Georges Cuvier Zoologist. A Study in the History of Evolution Theory*, Cambridge (Mass.), 1964: chap. 1.

4. See, for example, for Edinburgh, John Walker's *Lectures on Geology* (ed. Harold W. Scott), Chicago, 1966.

5. A. -L. Millin, 1792: quoted in Daudin, *Classes zoologiques*, p. 9.

6. Georges Cuvier, *Le Règne animal distribué d'après son Organisation pour servir de Base à l'Histoire naturelle des Animaux et d'Introduction à l'Anatomie comparée*, Paris, 1817, 4 vols.

7. See Coleman, *Cuvier*, chap. 2.

8. Cuvier, 'Discours prononcé par le citoyen Cuvier, à l'ouverture du cours d'Anatomie comparée qu'il fait au Muséum national d'histoire naturelle, pour le citoyen Mertrud', *Magasin encyclopédique*, 1re année, vol. 5, pp. 145–155 (an 4: 1795).

9. See Coleman, *Cuvier*, chap. 3, 4. Also E. S. Russell, *Form and Function. A Contribution to the History of animal Morphology*, London, 1916: chap. 3.

10. G. Cuvier, 'Notice sur le squelette d'une très-grande espèce de Quadrupède inconnue jusqu'à présent, trouvé au Paraquay, et déposé au Cabinet d'Histoire naturelle de Madrid', *Magasin encyclopédique*, 2me année, vol. 1, pp. 303–310. (1796).

11. See John C. Greene, *The Death of Adam. Evolution and its Impact on Western Thought*, Ames, Iowa, 1959: chap. 4.

12. Cuvier, 'Mémoire sur les espèces d'elephans vivantes et fossiles', *Mémoires de l'Institut national des Sciences et Arts, Sciences mathematiques et physiques*, vol. 2, *Mémoires*, pp. 1–22. (1799).

13. J. A. De Luc, *Lettres physiques et morales sur l'Histoire de la Terre et de l'Homme addressées à la Reine de la Grande Bretagne*, La Haye and Paris, 1779; 'Letters to Dr. James Hutton, F.R.S., Edinburgh, on his Theory of the Earth', *Monthly Review or Literary Journal*, vol. 2, pp. 206–227, 582–601; vol. 3, pp. 573–586 (1790); vol. 5, pp. 564–585 (1791); *Lettres sur l'Histoire physique de la Terre, addressées à M. le Professeur Blumenbach, renfermant de nouvelles Preuves géologiques et historiques de la Mission divine de Moyse*, Paris, an 6–1798.

14. See R. Hooykaas, *Natural Law and Divine Miracle. A historical-critical study of the principle of uniformity in geology, biology and theology*, Leiden, 1959 (2nd impression, 1964, under title *The Principle of Uniformity*); *Continuité et Discontinuité en Géologie et Biologie*, Paris, 1970.

15. G. Cuvier, 'Extrait d'un ouvrage sur les espèces de quadrupèdes dont on a trouvé les ossemens dans l'intérieur de la terre', *Journal de Physique, de Chimie et d'Histoire naturelle*, vol. 52, pp. 253–267 (an 9: 1801).

16. See Charles Gillispie, *Genesis and Geology. A Study in the Relations of Scientific Thought, Natural Theology, and Social Opinion in Great Britain, 1790–1850*, Cambridge (Mass.), 1951; Milton Millhauser, 'The Scriptural Geologists. An Episode in the History of Opinion', *Osiris*, vol. 11, pp. 65–86 (1954).

17. See Déodat de Dolomieu, 'Mémoire sur les pierres composées et sur les roches', *Observations sur la Physique, sur l'Histoire naturelle et sur les Arts*, vol. 39, pp. 374–407 (1791); vol. 40, pp. 41–62, 203–218, 372–403 (1791): esp. footnote on

pp. 41–3; Kenneth L. Taylor, 'The Geology of Déodat de Dolomieu', *Actes du XIIme Congrès international d'Histoire des Sciences*, vol. 7, pp. 49–53 (1971).

18. G. Cuvier, 'Sur les Elephans vivans et fossiles', *Annales du Muséum national d'Histoire naturelle*, vol. 8, pp. 1–58, 93–155, 249–269 (1806).

19. G. Cuvier, 'Mémoire sur le squelette presque entier d'un petit quadrupède du genre de Sarigues, trouvé dans le pierre à plâtre des environs de Paris', *Annales du Muséum national d'Histoire naturelle*, vol. 5, pp. 277–292 (1804). See also Coleman, *Cuvier*, chap. 5, and Russell, *Form and Function*, chap. 3.

20. See note 15.

21. See C. C. Gillispie, 'The Formation of Lamarck's evolutionary theory', *Archives internationales d'Histoire des Sciences*, vol. 9, pp. 323–338 (1957). Also Daudin, *Classes zoologiques*, chap. 10.

22. See Lovejoy, *Great Chain of Being;* Henri Daudin, *De Linné à Jussieu. Méthodes de la classification et idée de série en botanique et en zoologie (1740–1790)*, Paris, [1926].

23. See Daudin, *Classes zoologiques*, p. 43ff.

24. J. B. Lamarck, 'Discours d'ouverture du Cours de Zoologie, donné dans le Muséum National d'Histoire Naturelle l'an 8 de la République', *in: Systême des Animaux sans Vertèbres . . .* , Paris, An 9–1801: pp. 1–48. Reprinted in *Bulletin scientifique de la France et de la Belgique*, vol. 40, pp. 443–597 (1907); for a modern English translation, see D. R. Newth, 'Lamarck in 1800: a lecture on the invertebrate animals, and a note on fossils . . . ', *Annals of Science*, vol. 8, pp. 229–254 (1952).

25. J. B. Lamarck, *Hydrogéologie ou Récherches sur l'influence qu'ont les eaux sur la surface du globe terrestre; sur les causes de l'existence du bassin des mers, de son déplacement et de son transport successif sur les differens points de la surface de ce globe; enfin sur les changemens que les corps vivans exercent sur la nature et l'état de cette surface*, Paris, an 9 [1802]. For a modern English edition, see Albert V. Carozzi, *Hydrogeology by J. B. Lamarck*, Urbana (Illinois), 1964.

26. James Hutton, 'Theory of the Earth; or an investigation of the laws discernible in the composition, dissolution and restoration of land upon the globe', *Transactions of the Royal Society of Edinburgh*, vol. 1, part 2, pp. 209–304 (1788); reprinted as chap. 1 of *Theory of the Earth, with Proofs and Illustrations*, 2 vols., Edinburgh, 1795 (facsimile reprint, Weinheim and Codicote 1959). See also R. Hooykaas, 'James Hutton und die Ewigkeit der Welt', *Gesnerus*, vol. 23, pp. 55–66 (1966); and R. H. Dott, Jr., 'James Hutton and the Concept of a Dynamic Earth', *in* Schneer, *Toward a History of Geology*, pp. 122–141.

27. Desmarest, *Géographie physique: tome premier*, Paris, an 3 [1794]: pp. 732–782 (this work was published as part of the *Encyclopédie méthodique*). See also Rhoda Rappaport, 'Problems and sources in the history of geology, 1749–1810', *History of Science*, vol. 3, pp. 60–77 (1964).

28. Lamarck, 'Sur les fossiles', *in: Animaux sans Vertebres;* pp. 403–411. Translated by Newth (see note 24).

29. See note 12.

30. [Lacépède], 'Rapport des Professeurs du Muséum, sur les collections d'histoire naturelle rapportées d'Égypte, par E. Geoffroy', *Annales du Muséum national d'Histoire naturelle*, vol. 1, pp. 234–241 (1802).

31. John Playfair, *Illustrations of the Huttonian Theory of the Earth*, Edinburgh, 1802 (facsimile reprint, New York, 1956).

32. G. Cuvier, 'Mémoire sur l'Ibis des anciens Egyptiens', *Annales du Muséum national d'Histoire naturelle*, vol. 4, pp. 116–135 (1804).

33. Lamarck, 'Mémoires sur les fossiles des environs de Paris, comprenant la détermination des espèces qui appartiennent aux animaux marins sans vertèbres, et dont le plupart sont figurés dans la collection des velins du Muséum', *Annales du Muséum national d'Histoire naturelle*, vol. 1, pp. 299–312, 383–391, 474–9 (1802).

34. Lamarck, 'Considérations sur quelques faits applicable à la théorie du globe, observés par M. Peron dans son voyage aux Terres australes, et sur quelques questions géologiques qui naissent de la connoissance de ces faits', *Annales du Muséum national d'Histoire naturelle*, vol. 6, pp. 26–52 (1805).

35. J.-B.-P.-A. Lamarck, *Philosophie zoologique, ou Exposition des considérations relatives à l'histoire naturelle des animaux; à la diversité de leur organisation et des facultés qu'ils en obtiennent; aux causes physiques qui maintiennent en eux la vie et donnent lieu aux mouvemens qu'ils executent; enfin, à celles qui produisent, les unes le sentiment, et les autres l'intelligence de ceux qui en sont doués*, Paris, 1809. For a modern English translation, see J. B. Lamarck, *Zoological Philosophy* . . . (ed. Hugh Elliot), London, 1914 (reprinted New York, 1963). See also J. S. Wilkie, 'Buffon, Lamarck and Darwin: the originality of Darwin's theory of evolution', *in* P. R. Bell (ed.), *Darwin's Biological Work. Some Aspects Reconsidered*, Cambridge, 1959: chap. 6.

36. G. Cuvier, 'Sur les espèces d'animaux dont proviennent les os fossiles répandus dans la pierre à plâtre des environs de Paris', *Annales du Muséum national d'Histoire naturelle*, vol. 3, pp. 275–303, 364–387, 442–472; vol. 4, pp. 66–75; vol. 6, pp. 253–283; vol. 9, pp. 10–44, 89–102, 205–215, 272–282; vol. 12, pp. 271–284 (1804–8).

37. See A. G. Werner, *Kurze Klassifikation und Beschreibung der verschiedenen Gebirgsarten*, Dresden, 1787; Alexander M. Ospovat, 'Reflections on A. G. Werner's "Kurze Klassifikation" ', *in* Schneer, *Toward a History of Geology*, pp. 242–256; O. Wagenbreth, 'Abraham Gottlob Werners System der Geologie, Petrographie und Lagerstättenlehre', 'Werner-Schüler und Bergleute und ihre Bedeutung für die Geologie und den Bergbau des 19. Jahrhunderts', in *Abraham Gottlob Werner. Gedenkschrift aus Anlass der Wiederkehr seines Todestages nach 150 Jahren am 30. Juni 1967*, Leipzig, 1967: pp. 83–148, 163–178.

38. B. Faujas St-Fond, *Histoire naturelle de la Montagne de Saint-Pierre de Maestricht*, Paris, an 7 [1799].

39. G. Cuvier et A. Brongniart, 'Essai sur la géographie minéralogique des environs de Paris', *Journal des Mines*, vol. 23, pp. 421–458 (1808); *Mémoires de la Classe des Sciences mathématiques et physiques de l'Institut imperial de France*, an 1810, part 1, pp. 1–278 (1811).

40. Johann Carl Freiesleben, *Geognostischer Beytrag zur Kenntnis des Kupferschiefergebirges mit besonderer Hinsicht auf einen Theil der Graffschaft Mannsfeld und Thüringiens*, Freyberg, 1807–15.

41. See Joan M. Eyles, 'William Smith: some aspects of his life and work', *in* Schneer, *Toward a History of Geology*, pp. 142–158.

42. See Leroy E. Page, 'John Playfair and Huttonian Catastrophism', *Actes du XIe Congrès international d'Histoire des Sciences*, vol. 4, pp. 221–225 (1967).

43. G. Cuvier, *Recherches sur les ossemens fossiles de quadrupèdes, où l'on rétablit les charactères de plusieurs espèces d'animaux que les revolutions du globe paroissent avoir détruites*, 4 vols., Paris, 1812.

44. G. Cuvier, *Essay on the Theory of the Earth. With geological illustrations by Professor Jameson*, Edinburgh, 1813 (facsimile reprint, Farnborough, 1971), and later editions.

45. See Coleman, *Cuvier*, chap. 7.

46. William Buckland, *Vindiciae Geologicae; or, The connexion of geology with religion explained, in an inaugural lecture delivered before the University of Oxford, May 15, 1819, on the endowment of a Readership in Geology by His Royal Highness the Prince Regent*, Oxford, 1820.

47. James Hall, 'On the Revolutions of the Earth's Surface', *Transactions of the Royal Society of Edinburgh*, vol. 7, part 1, pp. 139–211 (1814).

48. William Buckland, *Reliquiae Diluvianae; or, Observations on the organic remains contained in caves, fissures, and diluvial gravel, and on other geological phenomena, attesting the action of an universal deluge*. London, 1823. See also its review [by W. H. Fitton], *Edinburgh Review*, vol. 39, pp. 196–234 (1823).

49. *Edinburgh philosophical Journal*, vol. 14, pp. 205–239 (1826). See Leroy E. Page, 'Diluvialism and its critics in Great Britain in the early nineteenth century', *in* Schneer, *Toward a History of Geology*, pp. 257–271.

50. Rhoda Rappaport, 'The Geological Atlas of Guettard, Lavoisier and Monnet: conflicting views of the nature of geology', *in* Schneer, *Toward a History of Geology*, pp. 272–287; Cuvier and Brongniart, see note 39; Freiesleben, see note 40; William Smith, *A Memoir to the Map and Delineation of the Strata of England and Wales with part of Scotland*, London, 1815; *Strata identified by organized fossils, containing prints on colored paper of the most characteristic specimens in each stratum*, London, 1816[–19].

51. Alexandre Brongniart, 'Sur les charactères zoologiques des Formations, avec application de ces charactères à la détermination de quelques terrains de Craie', *Annales des Mines*, vol. 6, pp. 537–572 (1821); *Mémoire sur les terrains de sédiment supérieurs calcaro-trappéens du Vicentin, et sur quelques terrains d'Italie, de France, d'Allemagne, etc., qui peuvent se rapporter à la même époque*, Paris, 1823.

52. Elie de Beaumont, 'Recherches sur quelque-unes des révolutions de la surface du globe, présentant différents exemples de coincidence entre le redressement des couches de certains systèmes de montagnes, et les changements soudains qui ont produit les lignes de démarcation qu'on observe entre certains étages consecutifs des terrains de sédiment', *Annales des Sciences naturelles*, vol. 18, pp. 5–25, 284–416 (1829); vol. 19, pp. 5–99, 177–240 (1830).

53. G. Cuvier, 'Sur un nouveau rapprochement à établir entre les classes qui composent le Règne animal', *Annales du Muséum national d'Histoire naturelle*, vol. 19, pp. 73–84 (1812); *Règne animal* (note 6). See also Coleman, *Cuvier*, chap. 4.

54. W. D. Conybeare & H. T. De La Beche, 'Notice of a discovery of a new fossil animal, forming a link between the Ichthyosaurus and the Crocodile; together with general remarks on the osteology of the Ichthyosaurus', *Transactions of the Geological Society of London*, 1st series, vol. 5, part 2, pp. 558–594 (1821); W. D. Conybeare, 'On the discovery of an almost perfect skeleton of the Plesiosaurus', *ibid.*, 2nd series, vol. 1, part 2, pp. 381–389 (1824); Gideon Mantell, 'Notice on the Iguanodon, a newly discovered fossil reptile, from the sandstone of Tilgate Forest, in Sussex', *Philosophical Transactions of the Royal Society of London*, vol. for 1825, pp. 179–186 (1825); Gideon A. Mantell, 'The geological age of reptiles', *Edinburgh new philosophical journal*, vol. 11, pp. 181–185 (1831).

55. See G. G. Simpson, *A Catalogue of the Mesozoic Mammalia in the Geological Department of the British Museum*, London, 1928: p. 3ff.

56. See John Lyon, 'The search for fossil Man: cinq personnages à la recherche du temps perdu', *Isis*, vol. 61, pp. 68–84 (1969).

57. E. F. von Schlotheim, *Beschreibung merkwürdiger Kräuter-Abdrücke und Pflanzen-Versteinerungen. Ein Beitrag zur Flora der Vorwelt*, Gotha, 1804.

58. Adolphe Brongniart, *Prodrome d'une histoire des végétaux fossils*, Paris, 1828; 'Considerations générales sur la nature de la végétation qui couvrait la surface de la terre aux diverses époques de formation de son écorce', *Annales der Sciences naturelles*, vol. 15, pp. 225–258 (1828). *Histoire des végétaux fossiles, ou recherches botaniques et géologiques sur les végétaux renfermés dans les diverses couches du globe*, 2 vols., Paris, 1828–1837.

59. Fourier, 'Remarques générales sur les températures du globe terrestre et des espaces planetaires', *Annales de Chimie et de Physique*, vol. 27, pp. 136–167 (1824); L. Cordier, 'Essai sur la température de l'intérieur de la terre', *Mémoires du Muséum d'Histoire naturelle*, vol. 15, pp. 161–244 (1827).

60. See M. J. S. Rudwick, 'Uniformity and Progression: reflections on the structure of geological theory in the age of Lyell', *in* D. H. D. Roller (ed.), *Perspectives*

in the History of Science and Technology, Norman (Oklahoma), 1971, pp. 209–227.

61. See Russell, *Form and Function*, chap. 5.

62. Geoffroy Saint-Hilaire, *Philosophie anatomique*, 2 vols, Paris, 1818–22. See also Théophile Cahn, *La vie et l'oeuvre d'Etienne Geoffroy Saint-Hilaire*, Paris, 1962.

63. Geoffroy Saint-Hilaire, 'Recherches sur l'organisation des gavials . . . et sur cette question, si les gavials (*Gavialis*), aujourd-hui répandus dans les parties orientales de l'Asie, descendent, par voie non interrompue de génération, des gavials antédiluviens . . . ', *Mémoires du Muséum d'Histoire naturelle*, vol. 12, pp. 97–155 (1825).

64. Geoffroy Saint-Hilaire, 'Mémoire où l'on propose de rechercher dans quels rapports de structure organique et de parenté sont entre eux les animaux des ages historiques, et vivant actuellement, et les espèces antédiluviennes et perdues', *Mémoires du Muséum d'Histoire naturelle*, vol. 17, pp. 209–229 (1828).

65. See Coleman, *Cuvier*, chap. 6; also J. Piveteau, 'Le debat entre Cuvier et Geoffroy Saint-Hilaire sur l'unité de plan et de composition', *Revue d'Histoire de Sciences*, vol. 3, pp. 343–363 (1950); but compare Frank Bourdier, 'Geoffroy Saint-Hilaire versus Cuvier: the campaign for paleontological evolution (1825–1838)', *in* Schneer, *Toward a History of Geology*, pp. 36–61.

66. See note 54.

Chapter Four

Uniformity and Progress

I

ON 11 January 1829, after three nights on a ship that was almost as uncomfortable as the primitive inns he had endured for two months in Sicily, Charles Lyell (1797–1875) landed in Naples. This marked the end of an expedition that had been decisive not only for Lyell's own intellectual development but also, through the enormous influence of his work, for the whole history of palaeontology. As he wrote the next day to his friend Roderick Murchison (1792–1871), who had accompanied him at an earlier stage of his travels, "we must preach up travelling . . . as the first, second and third requisites for a modern geologist, in the present adolescent stage of the science"[1].

The metaphor of adolescence was aptly chosen. Lyell's interpretation of geology—and that included palaeontology—was to dominate discussion within the science for the next three decades, whether by way of agreement or disagreement. However the science did not step ashore, Venus-like from the Bay of Naples, fully mature in the person Lyell. The publication of Lyell's *Principles of Geology* (1830–3) certainly initiated a lively debate on the most fundamental questions of the history of the Earth and of life itself, but Lyell's work did not create geology as a science where all had been speculation or undigested facts before. As we have seen, there was by 1830 a consensus of scientific opinion that embodied a synthesis of high scientific status and explanatory power. Lyell's work was important not because it suddenly replaced this with a more 'scientific' theory, but because it offered a radical challenge to the established synthesis, and forced scientists to re-examine its foundations. In the event, while some of Lyell's 'principles' were gradually assimilated and the consensus thereby modified, other elements of his thought were decisively rejected.

However, at the start of this period Lyell was correct in regarding the science as adolescent: it was growing rapidly, but it was still at a stage when new influences might alter fundamentally the direction of its development. One such influence, of which Lyell's expedition to the wilds of Sicily was at least a symptom, was the emergence of English geology as a major force within the international scientific community. To Lyell and to many of his friends, this was not without nationalistic overtones: "this year", he wrote to Murchison, "we have by our joint tour fathomed the depth and ascertained the shallowness of the geoloists of France and Italy," and it only remained to discover whether German science offered a stronger challenge to English supremacy. This new sense of corporate strength and confidence, while doubtless reflecting a similar mood in English thought in general at this time, derived its specific power, for men like Lyell and Murchison, principally from their participation in the Geological Society of London. This had been a vigorous and distinctly youthful body ever since its foundation in 1807. Lyell had become one of its Secretaries at the age of only 25—an office in which Murchison had later succeeded him—and was already a Vice-President at 31; while Murchison, also in his thirties, was now its Foreign Secretary, and even its President, William Fitton (1780–1861), was still in his forties. Such youthfulness perhaps contributed to the Society's rise to a pre-eminent position as a forum for geological debate in Europe. This was also favoured by the fact that it was the first learned society anywhere to be devoted specifically to the science; for after an almost mortal struggle with Sir Joseph Banks, the autocratic President of the Royal Society, it had won the right to exist independently of that still unreformed body[2].

The contrast between the Geological Society in London, and the Paris Museum which it was beginning to replace as the main centre for geological research, reflects accurately the contrast in the position of science in the two countries. Whereas the Museum and the Institute had been set up and liberally financed by the state, with support that continued through all the political upheavals of the subsequent years, the Geological Society, like other English learned societies and even the Royal Society itself, was a purely private body almost entirely dependent on its Fellows' subscriptions. In spite of the enormous economic importance of mineral resources to the world's first industrialised nation, state support for the science most likely to develop those resources remained almost non-existent[3].

Nevertheless, the attractions of geology as a science that was developing rapidly in spectacular directions, and moreover as a science in which almost anyone could hope to make exciting discoveries, was providing the Society with a steadily enlarging membership. Astronomy remained the aristocratic science in England, both socially and scientifically. However geology appealed to the tastes of the middle classes, as much as it suited their resources, for unlike astronomy its active pursuit required little more than a reasonable degree of leisure, the means for at least a little travelling, and a taste for the open air and the countryside. Lyell had trained and practised as a barrister, and when poor eyesight prevented him from following that career he was able, by supplementing his private means with earnings from authorship, to devote himself wholly to science. Murchison came from a similar family background of landowning gentry, and he had exchanged a military career for a life of fox-hunting when he was caught by the new enthusiasm for geology, a taste that he had the means to indulge as a full-time pursuit. Fitton and Mantell were professional doctors, Conybeare a clergyman, and most of the other prominent members of the Society came from similar social backgrounds. Of Englishmen with an extensive knowledge of geology, only the 'practical men' like William Smith and his associates were conspicuously absent from the Society, on account of their lower social class and inadequate financial means: the Society did in fact often draw on their knowledge, but it generally had to do so at second hand, through the personal contacts of its Fellows. In France (and Germany), by contrast, there was no rigid social or personal division between 'scientific' and 'technical' geology: the elder Brongniart for example was *Ingènieur en chef des Mines* before succeeding Haüy as professor of mineralogy at the Paris Museum, and he was at the same time deeply involved in ceramic technology as director of the state porcelain factory at Sèvres; while he and Cuvier had published their great memoir on Parisian stratigraphy in the *Journal des Mines* in the midst of articles on mining techniques. In England there was no such journal and no School of Mines either. At the two universities, on the other hand, there were now at least lectures in geology, even if they were strictly extra-curricular in nature: Buckland's at Oxford were matched in popularity

Fig. 4.1. Sedgwick's geological museum at Cambridge in 1842. Founded around the nucleus of John Woodward's earlier collection, it was enormously enlarged during Sedgwick's long tenure of the Woodwardian Chair of Geology. Note the skeleton of the Irish 'elk' in the centre—a crucial test-case in the 'diluvial' debates of the early nineteenth century (see Fig. 4.2).

by those of Adam Sedgwick (1785–1873), the first really active occupant of the Chair that had been founded at Cambridge by the will of John Woodward nearly a century before (Fig. 4.1).

Lyell had attended Buckland's lectures for three years running while an undergraduate at Oxford; and his sustained campaign against Buckland's diluvial theory and all that it seemed to him to represent must be seen against that background. In part it was a reaction, common enough in a brilliant pupil, against his teacher's favourite theory; in part a rejection of what he felt to be his teacher's slapdash methods of fieldwork; but most fundamentally it was a repudiation of Buckland's attempts to use geology in the service of a reconciliation with scripture. Lyell himself was not anti-religious; but he felt that the status of geology as a respectable science was threatened by Buckland's work. The market for popular books on geology was being met in England by a growing number of works that attempted to use geology to bolster a highly literalistic interpretation of scripture[4]; and while these were scientifically worthless, and recognised as such by men of science, Lyell felt that Buckland was betraying geology by lending the scientific authority of his position to this same endeavour to make geology conform with scripture. This feeling helps to account for the vehemence with which Lyell attacked the diluvial theory, and for his delight at any piece of evidence that could be brought into service to confound it. However, the momentum of his attack carried him far beyond a refutation of Buckland's distinctive variety of diluvialism into a far more radical critique of the whole synthesis that has been described in the previous chapter.

II

While Lyell was still somewhere in Sicily, Murchison had read to the Geological Society a paper that summarised some of their joint work in France and at the same time opened Lyell's campaign against the diluvial theory[5]. Although fossils played only a subordinate role in the argument, the paper illustrates a method of interpretation that was to have far-reaching implications for the understanding of the history of life. Lyell was historically correct in introducing his paper (the inspiration for it was almost certainly his, not Murchison's) as merely a reiteration of a theory "long ago announced" by Playfair and others, set in contrast to "the opposite opinions of De Luc and his school"—that is, of Buckland and his Oxford colleagues. The vulnerable point at which Lyell chose to attack diluvialism was in its interpretation of valleys. Playfair had used the slow excavation of valleys, by the streams now flowing in them, as a paradigm example of his quasi-mathematical method of reasoning in geology: by 'integrating' the small effects observable on the human time-scale, even very large effects could be produced on the geological time-scale. However to many of Playfair's contemporaries this interpretation had lacked persuasive power, not because they failed to accept the vast periods of time it demanded, but principally because the largest and most spectacular valleys in mountain regions were those in which the present streams were manifestly unrelated to the valley forms (the modern explanation of this is that such valleys have been radically modified by glacial action). The inadequacy of present streams to explain the excavation of valleys had therefore become a major example of the inadequacy of Playfair's thoroughgoing actualistic geology, and so, conversely, it had also been used as major evidence of the need to postulate some kind of diluvial event. The actualistic policy of using the present to interpret the past was certainly not rejected: as we have seen, it was taken by Cuvier himself to be the very feature that distinguished the geology of his own time from the speculative 'systems' of his forerunners. All that was disputed was the extent to which the present was an *adequate* key to the past.

This question had certainly not been forgotten or regarded as settled in the years before Lyell entered the debate. In the year that peace returned to Europe, Playfair's *Illustrations* were brought more widely to the attention of continental geologists by a French translation that very fairly coupled Playfair's work with one of the best critical evaluations that had been made of it when it was first published[6]. Interest in the question of the adequacy of 'actual causes' (that is, presently acting geological processes) was such that three years later (1818) the Royal Society of Sciences at Göttingen, at Blumenbach's suggestion, offered a prize for an essay "on the changes in the Earth's surface that can be established in history, and the application which can be made of this information in the investigation of the revolutions of the Earth that lie beyond the domain of history". The prize was won by the diplomat and amateur geologist Karl von Hoff (1771–1837), with a massive compilation of all the historical evidence for changes in physical geography through the processes of erosion and deposition, volcanoes and earthquakes, and so on[7]. The effect of this work was to show that these processes had been much more powerful, even within the geologically short span of human history, than had hitherto been supposed. This suggested that much more of the evidence of pre-human geological periods could be accounted for by their action, without any need to assume, as Cuvier had prematurely done, that the most drastic events had no parallel at all at the present day. It did not follow necessarily that there had been no such events, but only that they were likely to have been similar in kind, if not in degree, to those observed in human history: James Hall's suggestion that revolutions were caused by large-scale tidal waves is a good example of this point.

However, von Hoff's demonstration of the power of 'actual causes' could also be used to eliminate revolutions altogether, if it was combined with a Playfairian emphasis on the effectiveness of even slow processes over long periods of time. A major step in this direction was taken by another amateur geologist, George Poulett Scrope (1797–1876), Member of Parliament, in his studies of volcanoes[8]. Scrope's underlying motives for wishing to refute diluvialism were the same as Lyell's: Buckland's theory threatened the integrity and status of geological science not only by attempting a futile concordance with scripture but also because, as he said, "it stops further enquiry". In other words, to assume that no present process can account for past effects is positively *anti*-heuristic, because it discourages any search for a present process that might in fact be adequate. Scrope applied the opposite, that is actualistic, policy—"the only legitimate path of geological reasoning"—to the interpretation of the celebrated extinct volcanoes of central France; but his demonstration that they differed in no essential way from presently active volcanoes led him incidentally to a similar Playfairian conclusion about the excavation of valleys. In this area (which, fortunately for him, had been largely outside the range of glacial action), he was able to show that the occasional extrusion of lava-flows down the valleys had provided evidence of "several distinct steps in the process of excavation", as opposed to the diluvialists' sharp distinction between the pre- and post-diluvial states of the country (see Fig. 4.3). Gradual excavation by the streams themselves was thus adequate to explain the valleys—"with an unlimited allowance of time". This conclusion emphasised the overriding importance of *time* in geology: "every step we take in its pursuit", he wrote, "forces us to make almost unlimited drafts upon antiquity"—a banking metaphor that came appropriately from a political economist.

Scrope felt that he had heard Nature's refrain of "Time—Time!—Time!" echoing from all his geological observations. However as we have seen, it was not primarily an inadequate appreciation of the time-scale that made most gelogists believe that some kind of drastic revolutions had punctuated Earth-history. The effects themselves seemed to cry out for explanation in terms of sudden events, but between these revolutions extremely long periods of time were generally assumed. Nevertheless there was perhaps a gap between the rational acceptance of a vast time-scale and the imaginative appreciation of what such a time-scale might imply. It may be that this imaginative barrier could only have been broken by men with powerful ulterior motives for wishing to exploit a vast time-scale to the full. Certainly Scrope, followed by Lyell, realised that the immensity of time could be used as an explanatory device to eliminate the alleged 'catastrophe' that they abhorred: with "an unlimited allowance of time", ordinary geological processes were capable of producing even the largest effects, and apparently sudden and large-scale events could be 'ironed out' into slow and gradual changes no different from those that von Hoff had collected from historic records.

Scrope himself had no argument with Cuvierian revolutions as such, so long as they were interpreted in terms of present processes. Indeed he positively believed there had been occasional "partial [that is, localised] crises of excessive turbulence" at earlier periods, such as the "*paroxysmal* catastrophe" that had elevated the Alps; but even such events were explicable, he believed, since "rare combinations of circumstances sometimes gave a prodigious force" to ordinary processes. Scrope's whole theory of vulcanism was in fact based on the contemporary model of a directionally cooling Earth, and like von Hoff he took it for granted that the force of volcanic and seismic processes had diminished progressively in the course of time. He also took for granted the parallel 'progress' of life, though he was only peripherally concerned with this.

Scrope's attack on the geological aspect of the diluvial theory was neatly complemented at the same period by Fleming's attack on its biological basis[9]. The two attacks were closely parallel in character: Fleming sought to show that the extinction of the 'diluvial' fauna, like the excavation of valleys, had occurred not suddenly but gradually over a long period of time; and like Scrope he made the Playfairian actualistic method the explicit basis of his argument. Their motives were also similar: as we have seen, Fleming found Buckland's deluge as repugnant on scientific as it was on religious grounds. Like Scrope again, Fleming took it for granted that both the Earth and life itself had had a history of directional character, though he questioned the validity of the inference from fossils that the Earth had been hotter in the past. However his main concern was to use his knowledge of animal ecology and geography to show that the 'diluvial' species could have become extinct gradually by the piecemeal reduction and ultimate elimination of *local* populations. In support of this hypothesis he could point to the actualistic analogy of the effects of Man's activities within historic times: many species had become locally "extirpated" in Britain within the past centuries, and a few even totally extinct. Even without the influence of Man, the ordinary physical changes in local environments inevitably affected the ecology of all the species present, making some more abundant and other less, so that in the course of time it seemed likely that whole species would be lost.

Fleming's hypothesis of piecemeal extinction was also supported by the fact that in the bone-bearing gravels the remains of extinct species were often mixed with those of species still surviving; and this, he pointed out, "could be regarded as the death-blow of the diluvian hypothesis". If the alleged universal deluge had annihilated some species, why not all? Furthermore, at least one of the more spectacular extinct species, the Irish 'elk', was now thought to have survived into 'post-diluvial' times, for its remains were found in peat-bogs overlying the 'diluvial' boulder-clay, and it had apparently co-existed with Man (Fig. 4.2). If one member of this fauna had become extinct by causes other than Buckland's deluge, why not the others too? Fleming concluded that the evidence Buckland cited from his cave research simply did not warrant the inferences he had drawn from it; and that the 'diluvial' species—or some of them— had become extinct gradually and in piecemeal fashion as a result of the hunting activities of early Man, accelerated by natural environmental changes. Fleming's emphasis on the influence of Man made his hypothesis primarily a stick for beating Buckland, as it was designed to be; but his comments on the analogy between primitive Man and other predacious animals made it available as a more general explanation for extinction throughout Earth-history.

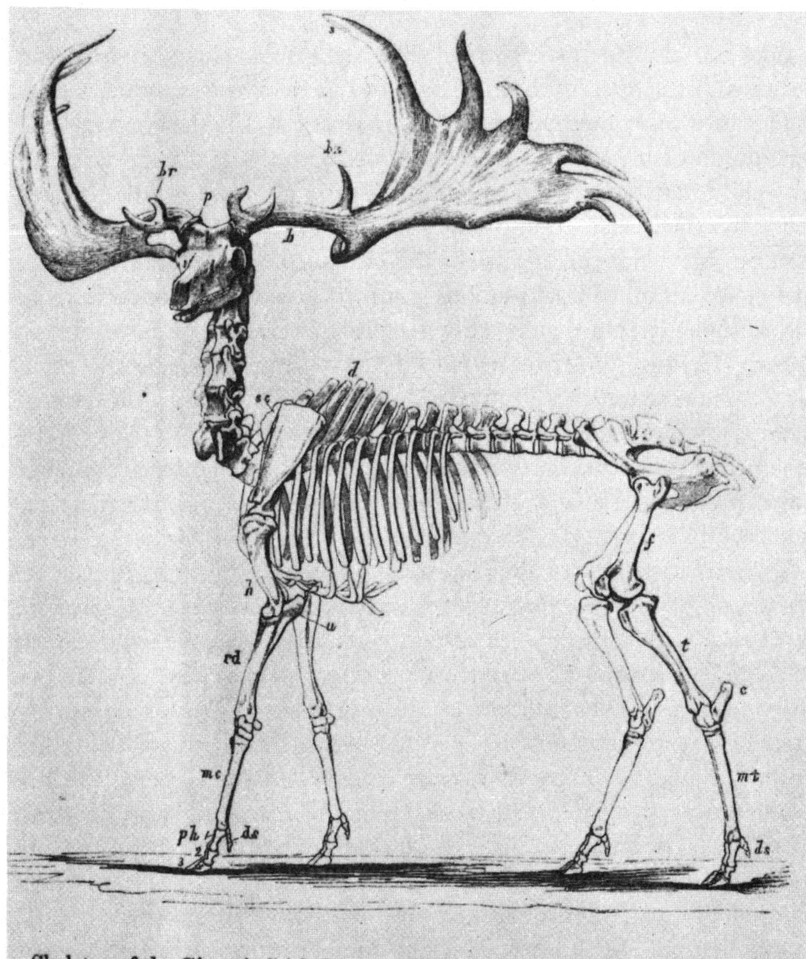

Skeleton of the Gigantic Irish Deer. Height to summit of antlers, 10 feet 4 inches.

Fig. 4.2. Richard Owen's reconstruction of the skeleton of the Irish 'Elk' (1846)[49]*. This was the most controversial of the 'diluvial' fauna of extinct mammals, because there was some evidence that it might have survived into post-'diluvial' times and have been exterminated by early Man. This interpretation—which was disputed—threw doubt on the 'catastrophic' extinction of the rest of the 'diluvial' fauna.*

III

It does not at all detract from Lyell's intellectual stature to point out how his distinctive approach to scientific problems was influenced by the ideas of his contemporaries and predecessors. On the contrary, the magnitude of his achievement is shown by the way that he was able to weld a vast range of ideas and observations, almost single-handed, into a synthesis that stood comparison with the one it opposed. Thus for example it was the breadth of his scientific interests and knowledge that enabled him to borrow both from the geologist Scrope and from the biologist Fleming, and to use their methods and conclusions in the service of a much wider synthesis of the two sciences. Like both of them he was motivated initially by the desire to refute Buckland's diluvial theory; and it was in an article for the *Quarterly Review*—one of a series designed to give the Tory intelligentsia a more 'enlightened' view of modern geology—that he first began to develop this synthesis clearly in public[10].

The article was ostensibly a review of Scrope's memoir on central France, but Lyell developed Scrope's arguments in a biological direction. Lyell saw that central France not only provided valuable evidence for the "gradual operation" of physical processes such as valley erosion, but also offered a possibly critical test of Fleming's hypothesis for the equally gradual nature of extinction. For this was an area that had apparently never suffered any Cuvierian marine incursion since before the deposition of fresh-water Tertiary strata like those around Paris. Between Cuvier's Tertiary faunas and those of the present, Lyell suggested, lay the period in the history of life "over which the greatest obscurity still hangs". Scrope had shown that the occasional eruptions in central France, had in effect, preserved intact some portions of the land surface from many different points of time within this period; therefore, Lyell argued, it was "still more important" to study whatever terrestrial faunas might have been preserved by the same accidents. For however the fauna had changed over this period, it could not have been by Cuvierian revolutions: central France at least had evidently remained immune from whatever physical changes had occurred in such low-lying regions as the Paris area. If in fact biological change was as slow and gradual as geological, then this extension of Playfair's doctrine would have to be demonstrated primarily by bridging the apparent gap between the faunas of the last 'regular' strata—Cuvier and Brongniart's Tertiary strata around Paris, and their equivalents— and those of the present.

By the time Lyell set out to fill this "page of the history of animated beings hitherto almost a blank", he was clearly confident that Buckland's deluge could be refuted and the whole of geology re-written without the need to postulate any recent revolution. He had even sketched an outline for a popular book on geology that would propagate these ideas and counteract the baleful influence of the scientifically worthless 'scriptural geologists.' Following the example of Scrope and Fleming the alleged catastrophe would be 'ironed out', with an appropriate "draft upon antiquity", into a long succession of slow processes indistinguishable from those of the present day.

In central France Lyell found Scrope's evidence for the slow excavation of valleys overwhelmingly persuasive: but he also found that a rich fossil fauna, recently discovered by the local naturalists Croizet and Jobert, clearly pointed to a similar conclusion for the history of life. The fauna consisted of extinct species of genera still in existence—in fact the warmer 'diluvial' fauna of elephants, rhinoceros, hippopotamus, hyaenas, tigers and so on. However it was buried under volcanic materials in ancient river gravels high above the present valley floor (Fig. 4.3). Thus by Scrope's criteria the fauna was of immense antiquity; and this implied that there could have been time enough for it to be replaced by the present fauna as slowly and gradually as the valleys had been excavated[11].

However, any process of gradual faunal replacement raised the problem of the origin of the newer species. Lyell was well aware that Geoffroy had recently revived Lamarck's theory of transmutation; but although such a gradual process might have appealed to him as analogous to slow valley erosion, his biological viewpoint was too Cuvierian not to side with "those more numerous and more distinguished physiologists" who were "opposed to such a conjecture". He might have evaded the issue by arguing that the newer species had migrated into Western Europe as gradually as the 'diluvial' species had become extinct; but this would only have pushed the problem further back in time, and would have been as unsatisfactory as a general explanation of faunal change as Cuvier's similar argument had been. The only alternative open to him was to assume that whatever the *means* by which species originated, their origins were as piecemeal as their

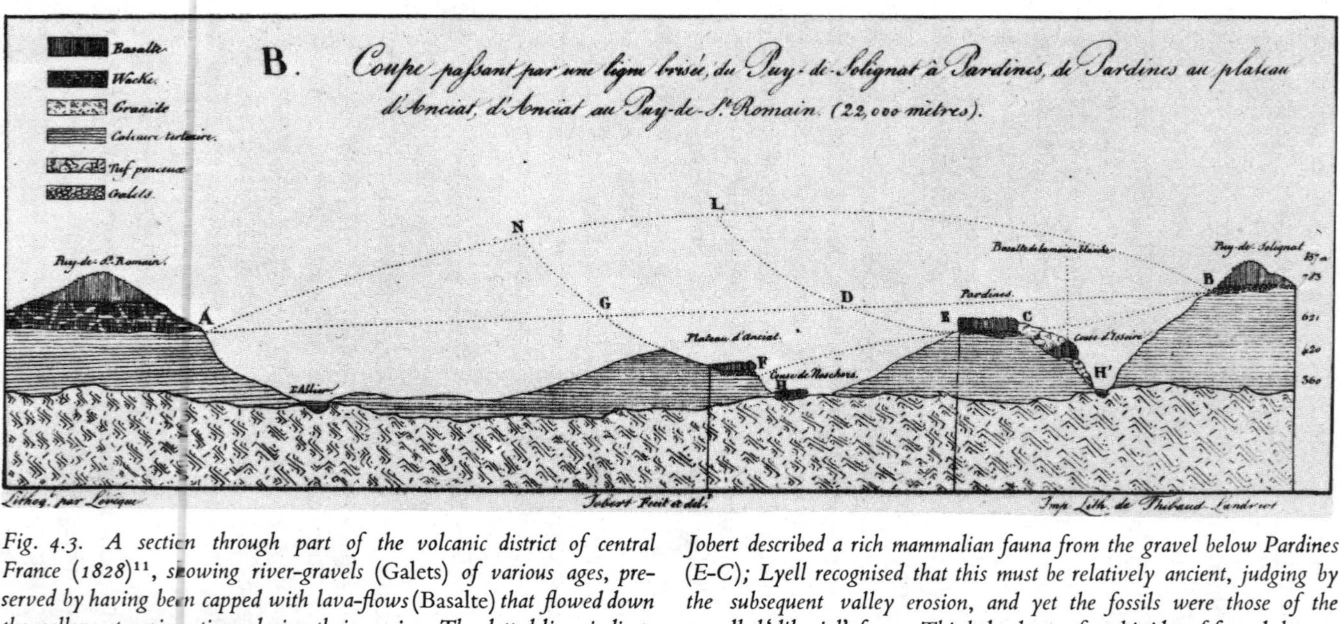

Fig. 4.3. A section through part of the volcanic district of central France (1828)[11], *showing river-gravels* (Galets) *of various ages, preserved by having been capped with lava-flows* (Basalte) *that flowed down the valleys at various times during their erosion. The dotted lines indicate reconstructions of the valley profiles at different stages. Croizet and Jobert described a rich mammalian fauna from the gravel below Pardines (E–C); Lyell recognised that this must be relatively ancient, judging by the subsequent valley erosion, and yet the fossils were those of the so-called 'diluvial' fauna. This helped to confirm his idea of faunal changes being relatively gradual and not 'catastrophic'.*

extinctions. This hypothesis preserved the reality of the species as a fundamental unit, as Cuvierian biological principles demanded; but at the same time it offered a model of faunal change that conformed to Playfairian geological principles.

With such a process of piecemeal production and extinction of individually stable species, the overall specific composition of a fauna would change just as gradually in the course of time as valleys were excavated or strata accumulated. This suggested to Lyell that just as Scrope had used degrees of valley excavation as a rough guide to the relative dates of successive lava-flows, so he himself might be able to use the specific composition of a fauna as a guide to the relative dates of strata. This possibility was of the utmost importance. If fossils could provide a 'natural chronometer' for Tertiary time (Lyell himself avoided using this phrase, since it had been used extensively by De Luc), it would not only help to fill up the 'blank page' palaeontologically but would also provide a time-scale on which to estimate the rates of geological processes too. In other words it might help to 'iron out' the appearance of sudden large-scale changes both biologically and geologically.

In Sicily this interpretation received support that made his expedition there the culmination of his travels. He confirmed to his own satisfaction that the huge volcano Etna had been built up gradually by the eruption of successive lava-flows and ash-falls, no different in magnitude from those recorded in history, on a time-scale that must have been immensely long by human standards. Yet he found that Etna stood on strata that were, by his reckoning, extremely recent by geological standards, since they contained a fauna composed almost exclusively of species still living in the Mediterranean (Fig. 4.4). Furthermore, these strata were greatly elevated in the centre of the island. From this he inferred not only that the strata themselves, and their fauna, neatly bridged the remaining gap between the present and the Subappenine strata of northern Italy, but also that there had been ample time for them to have been elevated by a process as gradual as the building up of the great cone of Etna. It followed that both biological and geological changes in the past could have occurred by processes as gradual in their action as those still in operation. It was superfluous to postulate any recent revolution in Earth-history: in Sicily at least the past could be shown to grade insensibly into the present[12].

"The results of my Sicilian expedition exceed my warmest expectations in the way of modern analogies", he wrote to Murchison on his

Fig. 4.4. Lyell's illustrations (1833)[14] of some of the mollusc shells characteristic of his 'Pliocene' period, based on the research of the French palaeontologist Deshayes. With its high proportion of species still extant, the Pliocene mollusc fauna was an essential part of Lyell's endeavour to connect the present with the geological past by a continuity of slow-acting processes.

return to Naples; but he had now arrived at a view of geology that went far beyond a mere vindication of the adequacy of 'actual causes' for the explanation of past Earth-history. The popular book he had planned earlier was now described in much more weighty terms: "it will endeavour to establish the *principle of reasoning* in the science"[13]. In fact, however, Lyell mentioned *two* radically distinct principles. The first was the actualistic principle that the processes which acted in the past were the same as those that act in the present: this of course was the policy which had guided Scrope and Fleming among others. Lyell's second principle was far more debatable in its implications: that these processes "never acted with different degrees of energy from that which they now exert". This conviction arose naturally enough out of Lyell's realisation, in an area of active vulcanism and frequent earthquakes, that present processes were much more powerful than was generally realised by geologists living in more stable areas; and that even the largest effects, such as the elevation of the Alps, could have been accomplished by processes of the same intensity, given the very long time-scale for which he now had good concrete evidence.

But by thus eliminating the necessity for *any* past events differing in character from those of the historic present, Lyell had followed Playfair, as Playfair had followed Hutton, into a position that conflicted irreconcilably with the views of his contemporaries. He had in effect denied the validity of the evidence for a directionally changing Earth and a progressively changing life: he had replaced the directional model of the history of the Earth and of life with a steady-state model similar to Hutton's. However, he was not pushed into this position simply by the inexorable logic of his own observations on 'actual causes'. On the contrary he found it inherently attractive, because like Hutton he regarded a steady-state system as both scientifically and theologically superior: scientifically, because it brought geology more closely into line with the prestige science of astronomy; and theologically, because a world in perpetual and harmonious balance could demonstrate the wisdon of the creation more effectively, he felt, than a world in which a temporal beginning and end could be envisaged.

IV

The first volume of Lyell's *Principles of Geology* was published little more than a year after his return to England; and although it was a further three years before all three volumes had appeared, Lyell himself stressed that the whole work formed a single connected argument[14]. His contemporaries were right to regard it as Huttonian not only in offering a steady-state model of Earth-history, with all the eternalistic overtones of such a model, but also in being as broad in its explanatory intentions as the 'systems' of Hutton and other eighteenth-century writers had been. Indeed Lyell himself freely used the word 'system' to describe his own work, despite its speculative overtones; and his work was quite explicitly an attempt to re-interpret the whole range of geological and palaeontological knowledge. In order to convince his readers of the validity of his re-interpretation he was bound to use as persuasive a style as he could. Sedgwick criticised him for using "the language of an advocate", but however much it ran counter to the supposedly 'Baconian' ideals on which the Geological Society had been founded, it was only natural that Lyell should have approached the subject with the advocacy of a trained barrister.

He opened his case with a retrospect of the history of the science, which he marshalled to support the need for a radical separation of geology from all questions of the interpretation of scripture. Having thus disposed of Buckland and the even less reputable 'scriptural geologists', he then went on to emphasise the importance of an adequate sense of time in geology; and he suggested how, without such a sense, a succession of the most ordinary events was bound to appear sudden and catastrophic. This was in effect a preliminary critique of Cuvierian revolutions. However, he then had to undermine the basic elements in the directional model of Earth-history; and here, as he realised, he faced the much "weightier objections" of a far wider range of his fellow-scientists. First he attacked the palaeontological arguments that the younger Brongniart and many others had used in support of a gradually cooling Earth. Unlike Fleming he could no longer believe that the evidence of climatic change was wholly illusory; but he was able to explain it away, within the framework of a steady-state Earth, by drawing on the results of research in physical geography. These showed that local climate was as much dependent on geographical circumstances of landmasses, winds and ocean currents, as on mere latitude. If, therefore, the physical geography of a given region had changed profoundly, by the action of geological processes, that region could have had many different climates in the course of time. Thus northern Europe might indeed have been tropical in the Coal period, as Brongniart inferred; yet this could not be taken as evidence that the whole globe had been hotter at that time.

The second objection to a steady-state system which Lyell had to meet at the outset was of course the palaeontological evidence for the 'progress' of life. With great boldness—or rashness—he argued that this, considered by his contemporaries to be one of the most firmly established generalisations in palaeontology, was simply an illusion. The appearance of mammals, for example, later than the 'lower' vertebrates was simply due to the accidents of preservation. Most mammals, he argued, were terrestrial and therefore unlikely to be preserved; and even the mammals of Stonesfield could be re-interpreted as proof that there had been *some* mammals in Secondary time. Conversely he argued that the Secondary period *seemed* to be an age of reptiles simply because most of them were marine forms and therefore more liable to be preserved. Only for Man himself did Lyell accept the conventional interpretation; but by asserting that the recent origin of Man was a novel event only on the "moral" (that is mental) level, he was able to deny that this was any evidence for "a progressive system".

Having sown the seeds of doubt in the minds of those who thought a directional theory well established, the next stage of Lyell's case was to show that the geological and biological agencies of change operating in the present were quite powerful enough to have accomplished all the observed effects in the past, without any need to postulate *either* occasional events of greater intensity *or* a gradual overall diminution of intensity. For his analysis of geological processes he drew heavily on von Hoff's examples, but deployed them in such a way as to emphasise the *balance* of opposing processes. Thus the powers of gradual erosion were balanced by those of deposition; and those of volcanic eruptions and seismic elevation by those of seismic subsidence. In this way he could argue that the physical state of the Earth's surface was in dynamic stability: the features of physical geography were changing continuously, but the Earth as a whole could have remained essentially in the same state.

The same model of a dynamic balance had next to be applied to the phenomena of life. However, Lyell had first to establish that species could be regarded as real units for the purpose of estimating the changes in the world of life. He had to give an extended critique of Lamarck's evolutionary theory—the *Zoological Philosophy* had just been re-issued as a result of Geoffroy's revival of the question—because it placed in doubt the very foundations of his own work. He was inclined anyway to believe in the stability of species on Cuvierian and ecological grounds; but to have admitted that species were unreal, and organisms in a state of constant flux, would have undermined the validity of his 'natural chronometer', on which in turn his demonstration of the steady state of Earth-history was to be based. It was therefore essential that he should demonstrate that although individuals do vary in form and habits there are real limits to the variability of a species and no "indefinite capacity of varying from the original type". Having established "that species have a real existence in nature", he then applied his concept of a continuously changing physical geography to show that it was bound to lead to a perpetually shifting pattern of ecological circumstances. This in turn implied continuous changes in the local abundance of different species, and hence could lead from time to time (as Fleming had suggested) to the complete extinction of particular species.

However, if the piecemeal extinction of species were thus "part of the constant and regular course of nature", he had to consider next "whether there are any means provided for the repair of these losses". If the world of life, like its physical environment, was in a state of dynamic stability, the extinction of species would have to be balanced by a corresponding process of piecemeal production of new species. At this point the metaphysical motivation for his steady-state system took precedence even over his methodological principles: for he was unable to produce any actualistic evidence for such a process of species-production occurring in the present. He was obliged to get round the difficulty by asserting that such an infrequent event was unlikely ever to have been observed within the short span of human history—though this was precisely the argument that his opponents had used to explain sudden revolutions. Lyell later said privately that he had left it to be inferred that new species were "created" by some unknown process as *natural* and 'secondary' in character as the process of extinction (a common enough meaning of the word 'creation' at this period)[15]: certainly he implied that new species had been produced, just as old ones had become extinct, in piecemeal fashion within space and time. He also implied that compared to the whole time-span of a species its 'creation' was a relatively sudden event; and that once 'created' its organisation necessarily remained adaptively stable until, sooner or later, it became extinct. Thus although Lyell was forced to leave the exact nature of species 'creation' an unsolved puzzle, he could integrate its undoubted occurrence into a model of organic change that paralleled his model of geological change. Both the physical world and the world of life were in a state of dynamic stability, with opposing processes perpetually in balance, at least on a global scale.

This model, however, had been argued in terms of what *should* be the case, given the processes that can be observed in the present. It still remained to show that the positive evidence of the past confirmed that such *had* in fact been the case. The study of present processes was, Lyell said, "the alphabet and grammar of geology": but the point of learning these was to be able to decipher the 'language' in which the records of the past had been written by nature, and hence to be able to understand the history of the Earth and of life. To read these records aright, however, it was essential to learn the correct 'language'. In other words, Lyell had to show that the fossil record of the terrestrial vertebrates, on which the 'progressive' model of the history of life chiefly depended, was in fact highly unreliable, owing to their haphazard fossilisation; and he argued that marine molluscs, with their much more consistent chances of preservation, provided the most reliable indication of the rate and nature of organic change (Fig. 4.4).

Fossil molluscs, "the demotic character in which Nature has been pleased to write all her most curious documents"[16], thus became the basis for Lyell's reconstruction of the Tertiary period. (Presumably, following this Egyptological metaphor, Cuvier and his followers could be left with the more spectacular but also more problematical hieroglyphic language of the vertebrates). Having established the probability of a uniform overall *rate* of change in the organic world, the various Tertiary deposits could be dated quantitatively on his 'natural chronometer' according to the percentage of extant species that they contained. Lyell distinguished arbitrarily four different portions of Tertiary time for which deposits were known, but by implication there were vast stretches of time between these, unrepresented by any strata yet discovered (Fig. 4.5, A).

This in itself was important, for it emphasised the extremely fragmentary nature of the geological record and the vastness of geological time, and hence left ample room for the 'ironing out' of the apparently sudden mountain elevations and faunal discontinuities used by Élie de Beaumont. For example, Lyell interpreted the apparently abrupt faunal discontinuity between the youngest Secondary and oldest Tertiary strata (that is, between Chalk and Eocene) as the result of an

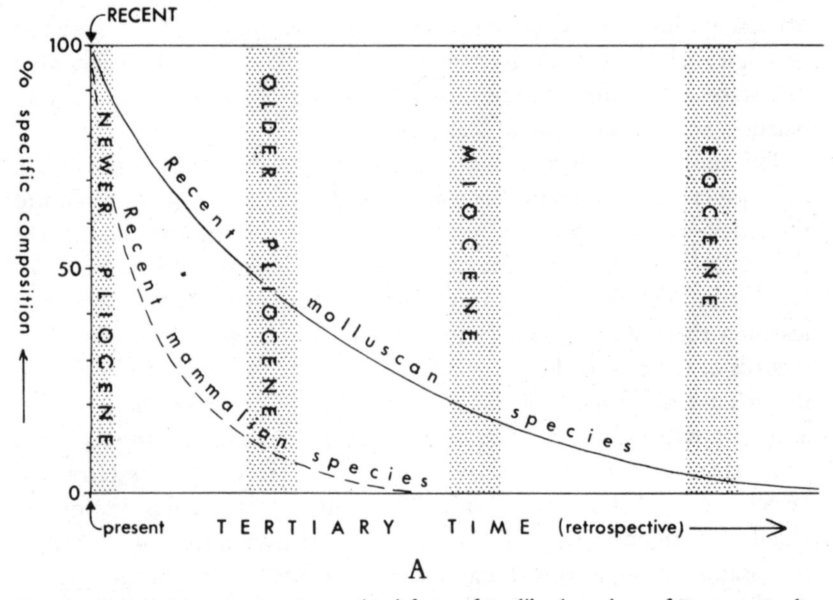

Fig. 4.5.A. An interpretation, in graphical form, of Lyell's chronology of Tertiary Earth-history, based on his hypothesis of a uniform overall rate of organic change. The temporal order of the known Tertiary formations ('Eocene', 'Miocene' etc.) and the unrecorded intervals between them are derived from Lyell's use of the percentage of extant molluscan species in each fauna as an index of its age (the use of a survivorship curve in this diagram is justified by Lyell's frequent reference to, and evident understanding of, the analysis of contemporary census returns). Lyell believed the rate of change of mammalian faunas was too rapid to be used in this way.

unrecorded time-span *longer* than that separating the Eocene from the present (Fig. 4.5, B); and this neatly eliminated the basis for postulating any sudden elevation of the Pyrenees during that interval.

Furthermore, however, for each of his preserved portions of Tertiary time Lyell collected evidence to show that the character and scale of geological processes had been no different from those of the present. This demonstrated, he believed, that his steady-state model was valid at least for the whole of Tertiary time; and a brief treatment of the Secondary strata suggested that it could be extended to them also. For the supposed Primary rocks, Lyell boldly tried to knock the bottom out of the 'progressive' record of life by suggesting that they were unfossiliferous not because they antedated the first appearance of life but because they had been altered by "metamorphism" and their fossils thereby destroyed. He could thus conclude his three-volume argument

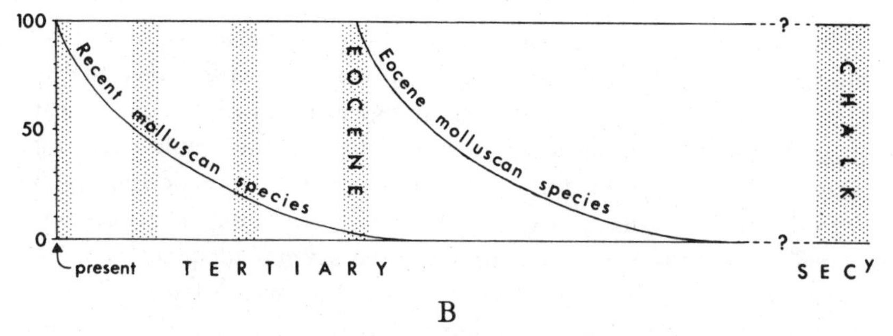

Fig. 4.5.B. A similar diagram to illustrate Lyell's inference that the faunal discontinuity between the oldest Tertiary strata and the youngest Secondary (Secy.) represented an unrecorded interval of time at least as long as the whole of the Tertiary, since there had been a complete turnover of molluscan species between them. This interpretation enabled him to eliminate the evidence for any 'catastrophic' event during this interval.

by asserting that as in space, so "in *time* also, the confines of the universe lie beyond the reach of mortal ken", and that geology, like astronomy, therefore provided limitless illustration of the ubiquity of wise design.

V

It is impossible to understand the influence of Lyell's *Principles* without recognising first that there were three distinct elements involved in his synthesis. The first of these was his actualistic *policy* for geological interpretation, which he stressed explicitly by subtitling the work 'an attempt to explain the former changes of the Earth's surface, by reference to causes now in operation'. However as we have seen, all Lyell's fellow-geologists without exception were agreed on the desirability of this policy. It is therefore not surprising that in general terms they welcomed Lyell's persuasive demonstration that very many of the phenomena of geology and palaeontology could reasonably be interpreted in actualistic terms: Conybeare, for example, felt that in this respect Lyell's work was "in itself sufficient to mark almost a new aera in the progress of our science"[17]. If they had reservations on this point, it was only because they doubted whether *all* the phenomena could be so explained, and because they felt that some (particularly those later interpreted as glacial in origin) might require explanation in terms of 'causes' different from those observable in operation.

Furthermore, however, it did not escape their notice that Lyell himself was singularly unsuccessful in applying the policy of actualism to that most puzzling of all palaeontological phenomena, the appearance of new species and new faunas and floras within geological time. This was the only phenomenon that seemed to some critics to be "utterly out of reach of any known laws of physiological action" and therefore a "remarkable exception" to the general adequacy of actual causes; and it was the only one for which Sedgwick's friend William Whewell (1794–1866) was even inclined to fall back on the view that a "direct manifestation of divine power" might be involved[18]. For all other phenomena, even those most striking in character, it was agreed that purely natural causes must be responsible: even if, like the sudden elevation of mountains, they could not be observed in operation at the present day, they were in principle explicable in terms of ordinary physico-chemical or biological 'laws'. As Conybeare commented, "no real philosopher" (that is, in modern terms, no scientist) doubted that the geological processes of the past had been identical at least in *kind* to those still in operation[19].

The second element in Lyell's argument was more debatable. This was his emphasis on the invariably *gradual* nature of geological and biological changes. In fact, Lyell's explanations tended to be 'gradual' in the original rather than in the modern sense of that word: many changes were taken to have occurred by a succession of small but sudden steps, rather than insensibly slowly. Thus mountains were elevated, in Lyell's view, by a long succession of sudden earth-quakes, and new faunas were formed by a long succession of 'creations' and extinctions of individual species. Lyell had no objection to sudden changes as such, so long as they did not overstep in intensity those that could be proved from historic records. Nevertheless, this very strict application of actualistic method obviously conflicted with the views of many of his contemporaries, who questioned whether the very short span of human history could legitimately be taken as an adequate guide to the whole of Earth-history. To Sedgwick, for example, Lyell's belief that present processes were "not only the type, but the measure of intensity" of all those of the past, seemed to be "a merely gratuitous hypothesis": it rested indeed on a basic philosophical confusion between "the immutable and primary laws of matter"—which had certainly been constant throughout time—and "the mutable results arising from their irregular combination"[20]. Thus Sedgwick contrasted the theoretical basis of

Lyell's work unfavourably with that of Élie de Beaumont, because the latter's conclusions seemed to him to arise from the phenomena themselves and not from *a priori* assumptions. Sedgwick was not forced into supporting Élie de Beaumont's theories on account of an inadequate sense of geological time: "of such an element as past time", he said, "we grudge no man the appropriation". However, the phenomena of violently folded rocks in mountain ranges and of abrupt faunal discontinuities in the record of strata seemed to him to indicate clearly that "long periods of comparative repose" had been punctuated by occasional "periods of extraordinary volcanic energy".

However it was the third element of Lyell's work that was most contentious, and on this he lost the support even of friends such as Scrope. In advocating a steady-state system for the history of the Earth and of life Lyell had to fly in the face of all the most firmly established physical and palaeontological evidence. As Sedgwick pointed out, this evidence clearly pointed to the gradual cooling of the globe, for which the astronomer John Herschel (1792–1871) had recently given a physically plausible explanation in terms of a diminishing eccentricity in the Earth's orbit[21]. Scrope, like Sedgwick, pointed out that the "general invariability" of the laws of nature would not in any way be violated by believing the Earth "to have passed through several progressive stages of existence"; whether the Earth's history had in fact been steady-state or directional in character should be decided from the evidence, and Scrope felt that the latter was clearly indicated. Lyell's 'metamorphic' explanation of exclusively 'Primary' rock-types such as gneiss and schist was ingenious, but Scrope felt it was more probable that they had been formed under physico-chemical conditions peculiar to the hot state of the Earth at that early period[22]. As for Lyell's dismissal of the fossil evidence for 'progression', many critics felt that this involved highly specious arguments that could hardly be sustained in the face of steadily accumulating palaeontological research.

In relation to the current research of the 1830s, therefore, Lyell's grand synthesis was as inopportune in its steady-state theory as it was timely in its emphasis on actualistic comparison with the present and on the relatively gradual nature of many geological processes. However, while this helps to explain the mixed reception of the *Principles*, it does not account fully for the enormous influence of Lyell's work, or for the fact that later in the century its publication seemed in retrospect to have marked a decisive step in the progress of science towards

Darwinism. In terms of the technical content of his work Lyell was in fact an 'odd man out' in the geological community of the 1830s, and he became in some respects even more so as the years passed. Yet within a broader intellectual debate his work became a paradigm example of the application of the 'principle of uniformity' to the world of nature. The discrepancy between these two levels of debate was due above all to Lyell's confusion—whether conscious or not—of several different meanings of 'uniformity'. In defending what Whewell dubbed inelegantly his "uniformitarianism", in other words his theory of a steady-state Earth-history without major 'castastrophes', Lyell believed he was holding the line against explanations that "stopped further inquiry" by placing the causation of past events beyond the realm of science, that is by attributing them to supra-natural agency. The 'uniformity' of his picture of Earth-history was thus hitched, not only to the methodological 'uniformity' involved in comparing past with present, but also to that fundamental metaphysical 'uniformity of nature' which alone could guarantee the intellectual autonomy of science itself.

Lyell's creative confusion about the 'uniformity of nature' ensured that successive editions of the *Principles* were read more widely and with greater enthusiasm than his elaborate technical argument might suggest. The popularity of the work was no doubt due in part to the general interest in geology and to Lyell's attractive and readable prose style; but above all the book was seen as formidable scientific support for a much broader intellectual position. The *Principles* not only undermined the credibility of attempts to 'reconcile' geology with scripture; more fundamentally Lyell's work was seen as vindicating an intellectual programme that admitted none but clear and distinct natural causes.

The reactions of the young Charles Darwin (1809–82), as he imbibed Lyell's work in the course of the *Beagle* voyage, are instructive in this respect. Darwin's mind was far from being a geological *tabula rasa* when he sailed for South America with the first volume of the *Principles* on board. He had had a 'crash course' in geology from perhaps the best teacher in England, Adam Sedgwick, and had learnt the techniques of the science with such good effect that even before his return from the voyage his observations were being treated with the greatest respect by far more experienced geologists. Yet precisely because the voyage gave him first-hand experience of grander phenomena than he had seen hitherto, it is only natural that he found Lyell's arguments persuasive. When he reached Chile and found ancient beaches with marine shells raised several hundred feet above the sea, he interpreted them at once in terms of the gradual (step-wise) elevation of the Andes; and while he was still there he actually saw his hypothesis confirmed by the disastrous 1835 earthquake, which elevated the coast a little further. His letters back to Cambridge (which, printed in his absence, formed his first scientific publication)[23] excited great interest, not least because his evidence for the gradual—and still continuing—elevation of the Andes threw doubt on Élie de Beaumont's theory of paroxysmal elevation. In that theory the sudden upheaval of the Andes had been specifically mentioned as a possible cause of the 'diluvial' tidal wave and of the last great episode of mass extinction. To re-interpret the Andes in Lyellian terms was therefore a suggestion with the most far-reaching implications.

Likewise it is symptomatic that as soon as Darwin saw the coral reefs fringing some of the Pacific islands he interpreted both them and true atolls as different *stages* in a single process of subsidence—a type of interpretation that he was later to apply to the problem of speciation. He even went further with Lyell than almost any other geologist of the time, when he interpreted the phenomena of oceanic islands not merely in terms of gradual depression or elevation, but also in terms of the *steady-state* movement of crustal masses. Some segments of the oceanic crust were, he believed, subsiding while others were simultaneously being elevated[24]. Yet although he thus applied a Lyellian steady-state theory to phenomena that even Lyell had not used in quite that way, Darwin did not follow Lyell in one crucial respect. He believed that Lyell was right to stress the fragmentary nature of the fossil record, but he never accepted Lyell's device for explaining away all the evidence of the progression of life. Unlike Lyell, Darwin therefore came to believe that while the Earth might have had a steady-state history, the history of life could have included the gradual development of 'higher' organisms (though without any intrinsically 'progressive' trend). In terms of the current theories, it might fairly be said that Darwin thus contrived to have the best of both worlds—a position that was not without significance for his work on the problem of species.

However it was not only new and impressionable recruits to the science, such as Darwin, who agreed with Lyell in his emphasis on gradual changes in the past history of the Earth and of life. Even those who disagreed fundamentally with his steady-state theory could still welcome his demonstration of the gradual nature of many changes, particularly of those in the more recent epochs of Earth-history. Indeed in this respect Lyell was only articulating, within a synthesis of exceptionally broad scope, a feeling that was becoming increasingly general among geologists. For example the relative dating of Tertiary deposits, and hence the subdivision of Tertiary time, by means of the relative proportions of extant molluscan species, is an outstanding case of simultaneous discovery. When Lyell stopped in Paris on his way back from Sicily, with this idea fully worked out in his mind, he found that Paul Deshayes (1796–1875), the greatest conchologist in Europe, had already come to the same conclusion; while the brilliant young palaeontologist Heinrich Georg Bronn (1800–62) of Heidelberg had already completed independently a similar but much more sophisticated statistical analysis of fossil faunas of all ages[25]. Behind the analyses of all three men there lay the belief that faunal changes were, at least generally, piecemeal and gradual in character, and not sudden or abrupt.

Such a belief was shared even by those who considered that geological processes might have been more intense in the earliest periods of Earth-history. Sedgwick, for example, who is often labelled a 'catastrophist', actually used his Presidential Address to the Geological Society just before the publication of Lyell's *Principles* as an occasion to stress the increasing evidence for the *gradual* nature of faunal transitions; and he and Murchison believed they had discovered, in the Gosau beds near Salzburg, a sample of those intermediate strata "which may hereafter close up the chasm" between the Secondary and Tertiary series—which was still one of the most puzzling 'chasms' known. Sedgwick even cited an example of an apparently abrupt discontinuity—that between the Coal Measures and the Lias in the West of England—specifically in order to show that this was *not* the result of "some short period of confusion" followed by "a new fiat of creative power", but that it was due merely to the local absence of intermediate strata and faunas known with increasing completeness elsewhere[26]. Sedgwick did not reject either the possibility of occasional paroxysmal episodes or the necessity of invoking "creative power" for the origin of individual species; indeed he positively praised Élie de Beaumont's work, and considered that ideas of the spontaneous generation or transmutation of species, with "their train of monstrous consequences", (that is, of materialism) were utterly unacceptable. But his position does illustrate how far a geologist with self-consciously empiricist principles could agree with Lyell's emphasis on gradual changes, while opposing any *a priori* "law of continuity" and reserving the right to postulate an occasional disruption of 'continuity' as and when the phenomena themselves seemed to demand it.

VI

It was only where Lyell's idiosyncratic concept of 'uniformity' led him to advocate a steady-state system, that Sedgwick felt it necessary to criticise his work with some asperity. This was justified, not only because Lyell's theory ignored all the contemporary geophysical evidence for the fluid origin of the earth, but also because current research was suggesting with increasing clarity that the 'progression' of life was far from being the illusion that Lyell asserted. It is ironical that the final and culminating volume of Lyell's *Principles*, in which his steady-state system was argued out in reference to the positive evidence of Earth-history, should have been published a few days after Murchison presented to the Geological Society a preliminary account of his epoch-making work on the Welsh Borderland: for this work, more than any other, was to undermine decisively the credibility of Lyell's steady-state theory.

To call it 'epoch-making' is not only an historically justifiable metaphor; it also is an appropriate pun, for it literally made new epochs in Earth-history out of what had hitherto been a confused complex of so-called Transition rocks. Werner had recognised long before that a 'transitional' series of rocks, containing sparse organic remains, ought to be intercalated between the totally unfossiliferous Primary rocks and the orderly and often highly fossiliferous Secondary strata; but the Transition strata were generally difficult to unravel, owing to folding and faulting, and they had remained poorly understood. Murchison, however, acting on a suggestion of Buckland's, had the good luck to find in 1831 an area in which the known succession of Secondary strata could be traced conformably downwards from the Old Red Sandstone into the Transition rocks. These Transition rocks proved, moreover, to be relatively easy to unravel: they were little altered, hardly more strongly folded than the Secondary strata of Smith's classic area, and above all they contained quite abundant and well-preserved fossils. It was the fossil contents of these strata that enabled Murchison to show, not only that Smith's methods could be extended into the Transition rocks, which others had despaired of ever reducing to order, but also that the Transition strata witnessed to an epoch in Earth-history as important as the 'age of mammals' that Cuvier's work had revealed.

Murchison chose to name the strata he studied "Silurian", after the tribe that had occupied the Welsh Borderland in Roman times. In defining the strata by reference to an area rather than a distinctive rock-type (cf. Carboniferous and Cretaceous) he was implying that fossils, and not lithology, should be the ultimate criterion for defining major series of strata. To some extent this was a judgement based on experience: he found, for example, that the well-known Wenlock Limestone could not be traced as a distinctive formation all the way into South Wales, and yet its fossils, or many of them, occurred there in the correct part of the succession. But it also reflected Murchison's conviction that the stratigraphical record contained a tolerably complete chronicle of the history of life, whereas the sediments in which fossils were buried were bound to owe their formation to much more localised causes. It was this belief that gave the Silurian fauna its outstanding theoretical importance.

The fauna was clearly marine in origin: there were corals, which in places even formed reef-like masses; there were trilobites—arthropods with large compound eyes, belonging to a group that appeared to be totally extinct—and there were abundant representatives of several groups, such as crinoids and brachiopods, which were known to survive, though only as rarities, in present seas, However, apart from some obscure fragments of fish in the very youngest strata, it was a fauna composed exclusively of invertebrates. Furthermore, although the sediments suggested deposition in warm shallow seas not far from land, there were no signs of any land plants. It seemed specious to explain away these facts in Lyellian manner as the results of differential non-preservation: a far more plausible and straight-forward interpretation was that the Silurian strata had been formed in an epoch prior to the first appearance of vertebrates or of terrestrial vegetation.

This conclusion would have been precarious if it had been based only on a study of the Welsh Borderland, but in fact its validity was strengthened year by year as Murchison himself traced the Silurian strata into South Wales, and as the Silurian fauna was recognised by other geologists on the Continent. After his first season's work,

Murchison described his preliminary results at the inaugural meeting of the British Association for the Advancement of Science. This was a 'pressure group' explicitly designed, as its title implied, to further the knowledge of science and the interests of scientists in a Britain that was still, officially and culturally, grossly neglectful of science and its potential applications. After a somewhat shaky start, the Association's meetings rapidly became a forum of scientific discussion at a very high level: its Geological Section, for example, was attended year after year by most of the best British geologists, by distinguished foreigners, and by scientists whose primary interests lay in other fields of science. Furthermore, by meeting each year in a different town away from London, the Association did much to counteract the tendency towards a metropolitan concentration of intellectual life and to recruit a wide range of amateur talent to the study of science—a recruitment that was particularly important for science like geology that depended so much on detailed local knowledge.

The further development of Murchison's work likewise reflects the contemporary scientific scene. After a second season of fieldwork he felt confident in announcing in more formal manner, this time as a paper to the Geological Society, his discovery of a distinctive fauna in a widespread series of Transition strata underlying the Old Red Sandstone. The name 'Silurian' was first used two years later, in one of the general scientific monthlies which at this perod (like *Nature* in our own day) provided a means of quick publication[27]. However the full publication of the work was delayed still longer, partly because he had to wait for his palaeontological collaborators to complete their detailed analysis of the Silurian fauna, and partly because the publication of engraved illustrations of fossils, stratigraphical sections and maps was a costly undertaking, and could only be achieved by the laborious process of soliciting pre-publication subscriptions from men of science and from the nobility and gentry of 'Siluria'.

The delay, however, was not without its compensations. By the time his sumptuous monograph on *The Silurian System* appeared at last in 1839[28], Murchison had good grounds for believing that the Silurian was no mere local series of strata, but a 'system' of probably world-wide validity. The Silurian fauna had been recognised by geologists all over Europe from Scandinavia to the Balkans; Darwin had collected Silurian fossils when the *Beagle* visited the Falkland Islands, and Herschel had sent similar specimens from Cape Colony where he was observing the stars of the southern hemisphere. Above all, huge spreads of almost flat-lying and unaltered strata with Silurian fossils were being reported from North America, notably by Timothy Conrad, the official palaeontologist to the New York State survey[29].

However the very success of the 'Silurian' concept could have raised theoretical problems. The relative uniformity of the Silurian fauna in such scattered parts of the globe was in striking contrast to the great regional and climatic diversity of Tertiary and living faunas. But, in the interpretation of this contrast "the geologist", wrote Murchison, "is not manacled by present conditions"—a criticism, of course, of Lyell's rigorist principles. If the Silurian fauna was uniform in composition, and hence implicitly in environment also, that was a fact to be explained, not merely explained away. "It would seem to be a fair inference", Murchison argued, "that however we may explain it, there must have *then* prevailed a generally more equable temperature". A satisfactory explanation for this was available in the current geophysical theory of a gradually cooling Earth: in the Silurian epoch, Murchison surmised, the greater influence of the Earth's internal heat could well have produced more uniform climatic conditions.

VII

The extension of the Silurian into a system of worldwide validity reinforced Murchison's belief that it represented an epoch of major significance in the history of life—an epoch that had antedated the appearance of vertebrates and land plants. This was important enough as confirmation of the 'progressive' character of the history of life; but it also had far-reaching economic implications. The major energy source for the rapidly expanding industries of the advanced nations was, of course, the coal of the Carboniferous strata. If the Silurian strata really dated from a period before there was any land vegetation, all areas of Silurian or earlier rocks could be confidently written off as potential sources of new coal seams: speculators who financed trial borings in such areas would be wasting their money.

This coincidence between theoretical and practical implications gave acute seriousness to a discovery that seemed at first to threaten the stability of Murchison's entire Silurian edifice. In 1834 Henry De La Beche (1796–1855) reported to the Geological Society that he had found fossil plants of Carboniferous species within the Transition rocks of Devon. His report was immediately questioned by both Murchison and Lyell, for it was incompatible with the theoretical outlook of either. De La Beche had failed to find any Silurian fossils in Devon, and therefore believed his slaty rocks were even older in date. But if so, Murchison argued, how could they contain the remains of terrestrial vegetation? The discovery was equally anomalous in Lyell's eyes; for although his steady-state theory enabled him to believe that land plants had existed even in pre-Silurian times, it was inconceivable to him that precisely the same *species* could have survived the vicissitudes of such a vast lapse of time and still be found in the coal strata of the Carboniferous period. To De la Beche these objections smacked of an apriorism in methodology which, if pursued consistently, would eliminate any real progress in knowledge of the history of life. How could that history ever be known, if the relative dates of strata were to be decided *a priori* from their fossil contents, rather than from observation of the actual order in which the strata had been formed? The coal plants occurred, he believed, in an integral part of the Transition strata: if that conflicted with theories about the nature of life in the Transition period, so much the worse for the theories[30].

But if the discovery was a disturbing anomaly for both Murchison and Lyell, their attempts to eliminate it also presented an immediate threat to De la Beche. Having lost his inherited income in the slump in the sugar trade that followed the abolition of slavery in the West Indies, he had returned from Jamaica faced with the distasteful necessity of earning a living; and he had been thankful to be appointed as a geologist to accompany the Ordnance Surveyors on their revision of the maps of south-west England. This was clearly seen—and not only by De la Beche himself—as a possible beginning of a much-needed involvement by the State in a scientific survey of the nation's mineral resources. That being so, nothing was more calculated to blight the emergence of a future Geological Survey, and to throw doubt on De la Beche's own professional competence to lead it, than Murchison's assertion that De la Beche had read the sequence of Devon strata upside down and that the fossil plants occurred in a huge unrecognised coalfield (admittedly without coal seams of great economic importance) overlying the real Transition rocks. In the event, the nascent Geological Survey survived even this attack, and De la Beche continued to insist that the problem was not yet solved, because the plant-bearing strata passed conformably into apparently ancient Transition rocks, without any intervening Old Red Sandstone or Silurian.

The ultimate solution of this dilemma was due, ironically, to a purely palaeontological hypothesis: William Lonsdale (1794–1871), the curator and librarian of the Geological Society, pointed out that the fossils that were being found in increasing numbers in the Transition rocks of Devon were intermediate in character between those of the Carboniferous Limestone and those of Murchison's Silurian System. This implied that the older rocks of Devon might be equivalent in date to the Old Red Sandstone; but they differed so much in appearance, and in the nature of their fossils, that Londsale's suggestion was not immediately accepted. Geologists were increasingly aware of the diversity of localised deposits, with distinctive fossils, that could be shown to date from the same epoch. However, the contrast between the Old Red Sandstone, with few fossils except strange-looking armour-plated fish (Fig. 4.6), and the slaty rocks of Devon, with limestone bands containing fossil corals and brachiopods, seemed too great to be explained in terms of differing environments at one and the same period of time.

Yet the validity of the 'Devonian System' was soon strengthened by Murchison's discovery, on a journey undertaken as soon as his Silurian book was safely published, that the Devonian fauna could be found on the Continent in the right position underlying the Carboniferous strata; and its equivalence with the Old Red Sandstone was finally clinched the following year (1840) when he found Old Red Sandstone fish and Devonian shells interbedded in the undisturbed strata of European Russia, where they were unambiguously overlain by Carboniferous strata and underlain by Silurian[31].

This prolonged and often heated controversy might seem to have been only of technical geological importance. But in fact it not only illustrates the entrenched theoretical commitments of all concerned; its ultimate resolution is also symptomatic of the way in which some of Lyell's principles were being absorbed by the scientific community as surely as others were being discarded. The acceptance of a Devonian System, intermediate—both temporally and palaeontologically—between the Silurian and the Carboniferous, reinforced the conception of the history of life as characterised by piecemeal change, and thus vindicated Lyell's emphasis on the gradual nature and intelligible causation of terrestrial processes in general. At the same time, however,

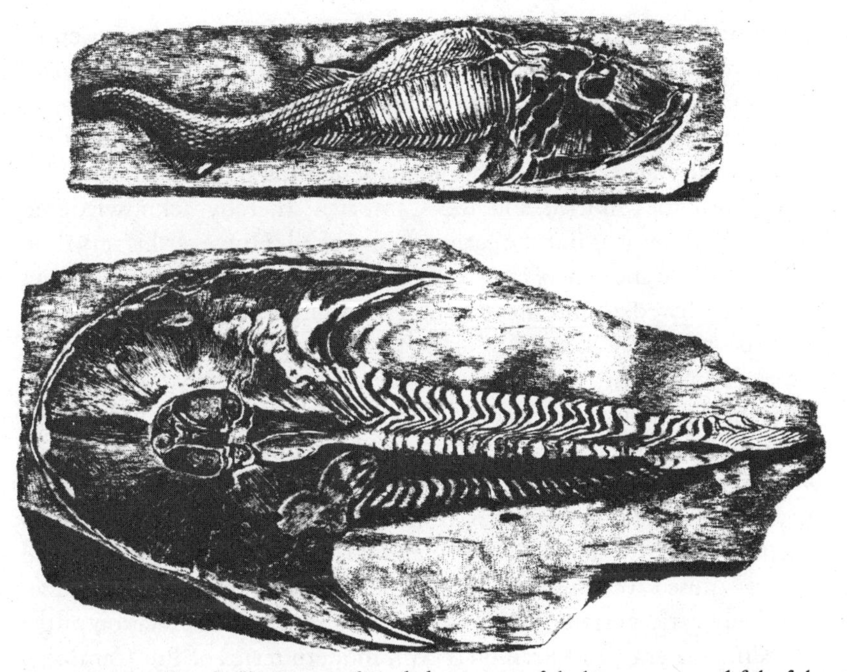

Fig. 4.6. Murchison's illustrations of Cephalaspis, *one of the bizarre armoured fish of the Old Red Sandstone (1839)*[28]. *These fish were the earliest vertebrates known; their virtual absence from the underlying Silurian strata reinforced the conception of the history of life as broadly 'progressive' in character.*

this research also reinforced the conception of the history of life as directional, and in a limited sense even progressive: there had been a distinctive temporal sequence of faunas and floras, in which the 'higher' or more complex forms seemed to have been appeared progressively.

With such a grand chronicle of the history of life coming to light it was natural that there should have been great interest in the question of what its earliest chapters might reveal. Was there any 'vestige of a beginning' to life itself, or had any such traces long been destroyed, as Lyell asserted, by the action of metamorphism? When Murchison began his work on the Welsh Borderland, Sedgwick had simultaneously attacked older and more complex Transition rocks in the heart of Wales (having the assistance of Darwin for one brief season before he left for the *Beagle* voyage). While Murchison defined a Silurian system, Sedgwick proposed a "Cambrian" beneath it[32]. Unfortunately for the future of their cooperation, Sedgwick could not point to any rich or distinctive Cambrian fauna, though fossils similar to Silurian forms certainly occurred in what he termed upper Cambrian strata, and even in the lower strata some traces of organic life could occasionally be found.

In his Silurian book Murchison suggested the term "Protozoic" to cover both the Silurian and the Cambrian, thereby acknowledging that the Silurian was not the earliest fauna of all. However his travels in Russia and Scandinavia later caused him to make more ambitious claims. In Sweden and around St Petersburg (Leningrad) he found what he termed a Lower Silurian fauna simply petering out when traced downwards; and in Sweden these strata lay directly on crystalline 'Primary' rocks such as schist and gneiss. The latter, which were totally unfossiliferous, he termed "Azoic"; and although he claimed that he was not thereby asserting dogmatically that no life had existed when they were formed, he evidently believed that to be the case. He agreed that they were similar to truly metamorphic rocks, but he believed this was caused by their formation under similarly hot conditions at an extremely early period of Earth-history. With this interpretation of the crystalline basement of Scandinavia (in modern terms of Pre-Cambrian age) as Azoic, the Silurian System became in Murchison's eyes still more important. It represented not only a major epoch before the first vertebrate life, but also the very first epoch with any record of life at all: "we can fearlessly assert", he concluded, "that the geological history or sequence of the earliest races of fossil animals is firmly established"[33].

Murchison's enlargement of the Silurian concept led to a long and bitter controversy with Sedgwick, and to their complete personal estrangement. Superficially it might seem to have been no more than a squabble about names: Sedgwick's 'Cambrian' strata turned out to overlap seriously with Murchison's 'Silurian', and Sedgwick was unable to point to any earlier distinctive fauna as the true Cambrian in his type area. More fundamentally, however, it was a controversy about the origin of life, for both men ardently desired the honour of showing that geology provided concrete evidence that life had indeed had a temporal beginning. Sedgwick's belief that the 'Cambrian' rocks of central Wales formed a pre-Silurian 'system' of major importance was eventually vindicated. A distinctive "Primordial" fauna of trilobites—the name indicates the theoretical importance attached to it—was discovered in the 1840s in Bohemia[34], and quickly recognised in Scandinavia too; but Murchison insisted on regarding it as the lowest subdivision of the Silurian. It was only later that the same fauna was discovered in Wales in Sedgwick's Cambrian strata, and so ultimately became the palaeontological basis for the Cambrian System of modern geology.

However, even the Primordial fauna failed to throw much light on the origin of life, indeed it made it more mysterious than ever. Whether it was classed as Silurian or Cambrian, there was no doubt that the Primordial fauna was underlain by strata with few or no traces of life; and this could not be explained away as the result of metamorphism, since in some cases (for example the Longmynd rocks of Shropshire) the sediments were no more altered than many fossil-bearing Silurian rocks. Furthermore, the fossil record did not begin with faint traces of primitive or rudimentary forms of life; it seemed to leap into existence with highly complex invertebrates (notably the trilobites) which clearly belonged to one or other of the still-existing Cuvierian classes. But if the nature of life's origin was more mysterious than ever, at least its temporal occurrence now seemed firmly established.

In 1840 Buckland summarised the trend of all this research with a vivid contemporary metaphor, when he said, "we are, as it were, extending the progressive operations of a general inclosure act over the great common field of geology"[35]. Both temporally and spatially, the earlier strata were being progressively reduced to order; behind the bewildering variety of local formations and the confusing effects of local tectonic disturbances, a unified scheme of major 'systems' was proving to be applicable in all parts of the globe. The empirical confirmation of the validity of this scheme strengthened the theoretical conclusion that could be drawn from it: the history of life, on the traces of which the scheme was founded, had followed in broad outlines an intelligible sequence of 'progressive' stages.

Although this progressive character could be illustrated from the history of plants and invertebrates, it was seen most clearly in the record of the vertebrates. Murchison had shown that the Silurian almost completely predated the first vestiges of vertebrate life; his collaborator Louis Agassiz (1807–1873), an ambitious Swiss naturalist who had worked briefly with Cuvier before the latter's death, showed that the subsequent Devonian period had been an age of diverse and often bizarre fish[36]. No higher vertebrates were yet known from the Carboniferous System; but some 'saurians' (that is, reptiles) had been found in overlying strata which Murchison in 1841 named the "Permian

System". This, then, formed the beginning of an 'age of reptiles' that extended through the rest of the Secondary strata, leading finally to the Tertiary as an 'age of mammals' and to the present as an 'age of Man'.

It is appropriate that it should have been William Smith's nephew, pupil and biographer, John Phillips (1800–1874), who proposed in 1841 the supreme divisions of geological time which summarised this sequence (Fig. 4.11). To emphasise the prime importance of the fossil evidence, the old divisions of Transition, Secondary and Tertiary could be replaced, he suggested, by Palaeozoic, Mesozoic and Cainozoic, the eras of ancient, middle and newer forms of life[37]. The record of life began, above the Azoic rocks, with the invertebrate faunas of the Silurian, and these were joined successively by fish, land plants and higher vertebrates before the Permian brought the Palaeozoic to a close. By Mesozoic time many of the Palaeozoic groups had waned in importance or, like the trilobites, had become totally extinct; the era was dominated by its reptiles (though a few primitive mammals had evidently been in existence), and, among its invertebrates, by such prolific molluscan groups as the ammonites and the belemnites. The Cainozoic showed a still greater approximation to the present, in its dominant land fauna of mammals, its dicotyledonous flora, and its marine fauna of 'modern' invertebrates and fish. Finally, in the very last scene of the geological drama, Man had appeared.

VIII

The research which has just been outlined covered a decade of intensive and brilliantly successful work. Murchison stands out as an obvious pivotal figure in its development; and as one whose enthusiasm for mixing with the titled nobility of Europe ultimately exceeded even his enthusiasm for geology, he was in no way averse to the fame and honours that it brought him. But he could never have achieved his grand perspective without the patient detailed work of a steadily growing international community of palaeontologists and geologists, who were publishing accurate identifications and descriptions of the fossils and strata of local areas all over the world.

Such scientists—the term coined by Whewell can at last be used without anachronism—were increasingly professional in outlook and in social position. The sciences of geology and palaeontology were now recognised generally as being of great cultural significance, for they provided an astonishingly novel perspective on the temporal place of man in nature. At the same time they were clearly of the highest importance economically, owing to their spectacular success in providing a rational basis for the discovery and exploitation of mineral resources. An increasing number of those who contributed to geological and palaeontological research were therefore professionals in the narrower sense: they were paid to do this work, in universities, museums, schools of mines and geological surveys. Even those who were, in this sense, amateurs, were imbued with a similar 'professional' spirit; and although this tended to lead to ever greater specialisation it did at least foster high standards of rigour and precision. Such standards were reinforced and maintained by the growing number of geological societies, whose refereeing procedures controlled the publication of specialist papers and whose medals, prizes and other honours served to reward work that conformed to these standards. The increasing specialisation of the science tended to remove its results further from the comprehension of the layman, and there was therefore a spate of popular and semi-popular works on the subject. Many of these were written by the most active 'professionals' themselves: Lyell for example brought out a small volume entitled *Elements of Geology* (1838), adapted from part of the *Principles*, and his friend Gideon Mantell published successful popular books that catered for the fashion for collecting fossils[38].

Although Lyell continued to assert that there was no valid evidence for the 'progression' of life, almost all other works on the subject, whether specialist or popular, emphasised that the fossil record was indeed broadly 'progressive', and that this could no longer be explained away as the result of imperfect or differential preservation. By the mid-1840s, the major outlines of the fossil record had thus been firmly established along lines that have survived with only minor modifications into mid-twentieth-century science. With such modernity apparent, it is natural to ask what causal explanations were brought forward to account for the 'progressiveness' of the history of life, or, more particularly, why an evolutionary explanation was not acceptable.

There were, of course, theological difficulties involved; but these were not, at least in scientific circles, primarily those of rejecting the concept of species-creation *ex nihilo*, Most scientists, many of them personally devout men, were quite prepared to accept a 'secondary' cause for the origin of new forms of life; indeed they were positively eager to do so, if a satisfactory natural cause could be discovered. That species, like other entities, should be 'governed' by God through the intermediary of 'natural laws' was, after all, simply in accord with their belief about the rest of the natural world. When speaking of the origin of new species, they commonly used vague and non-committal language: Murchison for example spoke of new groups being "brought into being"[39]. Even when, like Lyell himself, they used the verb 'created', it did not necessarily imply a belief that species had been formed from nothing by the direct activity of God: the language of 'creation' was a commonplace in science at this period, and tells us nothing about the means envisaged. When John Herschel referred to the origin of species as the "mystery of mysteries"[40] he certainly did not mean that it was a problem forever incapable of solution, but simply that it was proving extremely difficult to solve.

What was far more serious was the difficulty of conceiving any natural cause that would preserve the sense of the 'designfulness' of organisms. It is difficult for us to re-capture the intensity and pervasiveness of the belief at this period that the whole universe did indeed disclose the designing hand of God—whatever secondary means he might have employed to produce that effect. This belief was not confined to the formally religious: it penetrated the imagination of almost all thinking men and powerfully affected the kinds of scientific explanation they were prepared to entertain. As we have seen, organic design had been, ever since Ray, the most persuasive class of evidence tending to support this view; to suggest, therefore, that such design was illusory, and that the adaptive structures of animals and plants had arisen purely by chance, was for most men literally inconceivable.

Furthermore, the development of palaeontology had actually reinforced this traditional feeling about the natural world. In one of the best and most original of the Bridgewater Treatises of the 1830s, a series that was intended to refurbish the Paleyan tradition of natural theology from the latest discoveries of contemporary science, Buckland had shown how geology and palaeontology demonstrated the pervasiveness of design throughout Earth-history[41]. Design was not confined to the present state of the world, nor had the world emerged gradually from a less designful state: on the contrary, he argued, even the earliest known forms of life, such as the Silurian trilobites, showed the same remarkable correlation of structure and function as any living animal (Fig. 4.7). Buckland's work was immensely influential (it reached a wide readership among Continental scientists through its translation by Agassiz), for it reinforced persuasively the traditional theological interpretation of adaptation—a feature of the organic world that no respectable scientist could ignore.

For adaptation remained the crucial scientific issue at the heart of the problem of the origin of species. To invoke a Lamarckian 'tendency to progressive improvement' was not merely to regress to an outmoded *kind* of explanation, but also to fly in the face of the available evidence. Certainly there was a general overall 'progressiveness' about the fossil record; but in detail it did not show the gradual 'development' that a Lamarckian theory required. As Murchison's collaborator Edouard de Verneuil remarked in summarising the character of the Palaeozoic faunas, they clearly did *not* support the "old idea" (that is Lamarck's) that the first organisms were merely "nature's imperfect sketches" (*ébauches*)[42]. Only by questionable reasoning could it be asserted that the Silurian trilobites, for example, were in any sense 'less perfect' or 'simpler' or 'lower' than later arthropods, or the elaborately armoured Devonian fish than the later fish. Nor could Geoffroy's modification of Lamarck's theory serve any better: his use of monstrosities to account for the origin of new forms could hardly explain the universal occurrence of accurate adaptation; and to account for the astonishing complexity and beauty of adaptive structures by reference to chance abnormalities of development must have seemed almost repugnant and even perverse.

Furthermore the reality of the species as a discrete unit in nature seemed more firmly established than ever. Intra-specific variation, though an acknowledged phenomenon, seemed to have definite inherent limits; and the discrete character of species seemed to be confirmed by the evidence of fossils. The possibility of infinitesimally gradual shifts across inter-specific boundaries was therefore almost inconceivable (except to Darwin, when he privately queried Lyell's assertion of just this point); but that seemed to leave Geoffroy's saltatory trans-specific 'jumps' as the only—but unsatisfactory—alternative.

Finally, even if the origin of new species were to have been solved satisfactorily, a much more serious problem would still have had to be faced. It was one thing to speculate that a species might have changed—somehow—into a similar form within the same genus or even the same

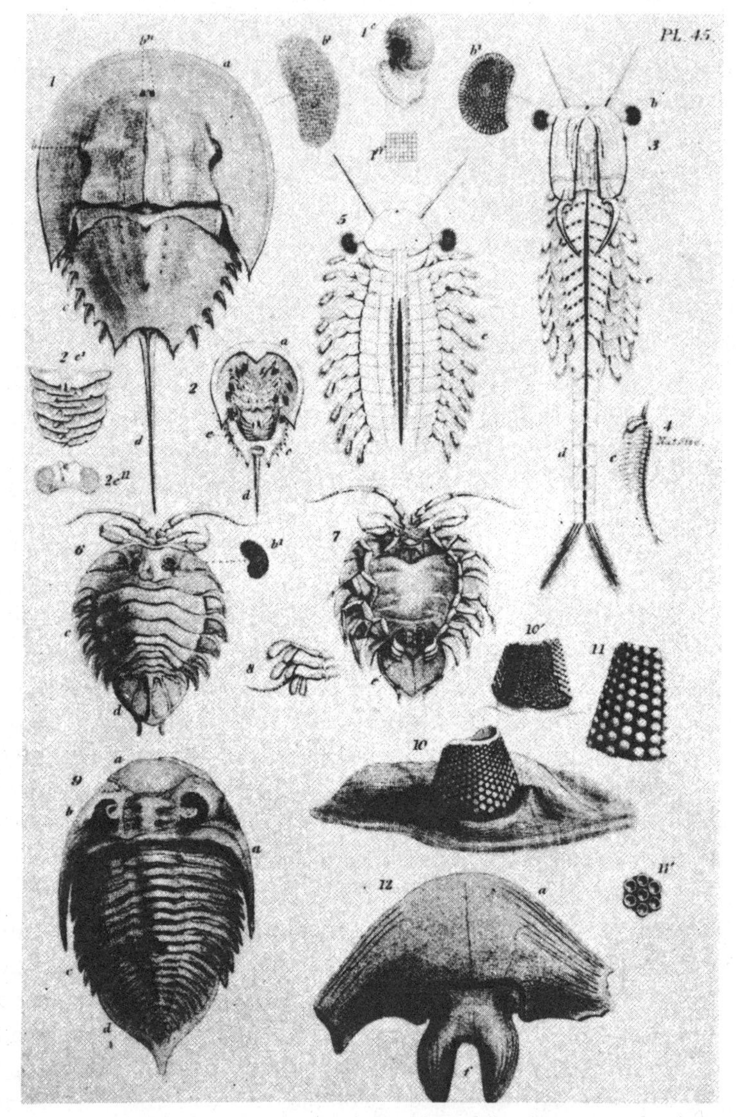

Fig. 4.7. Buckland's illustrations of the 'designful' adaptations of Silurian trilobites, with analogies from living arthropods (1836)[41]*. Figs. 9 – 11 show the compound eyes and all-round vision of the trilobite* Asaphus. *The xiphosuran (King-crab)* Limulus *(Figs. 1, 2) and the crustaceans* Branchipus *(Figs. 3 – 5) and* Serolis *(Figs. 6, 7), with their compound eyes, are shown for comparison. Buckland's analysis of trilobite adaptations was one of the most original of his examples of the designful nature of the organic world at even the earliest known period in the history of life.*

family; it was quite a different matter to suggest that the major groups of organisms (Cuvier's *embranchements*, or the phyla of modern taxonomy), with radically different anatomical construction, might have had a common origin. The validity of Cuvier's comparative anatomy, like his emphasis on the discrete nature of species, seemed to have been vindicated brilliantly by more recent palaeontological research. All the major groups of animals were as clearly distinct at their points of appearance in the fossil record as they were thereafter: traced back in time, they showed no signs whatever of converging towards a common ancestor.

IX

In the face of all these difficulties, it is not surprising that the one well-publicised attempt at an evolutionary theory in this period should have been scorned by the scientific community. The Scottish journalist Robert Chambers (1802–1871) entitled his anonymous book *Vestiges of the Natural History of Creation* (1844)[43]. The 'vestiges' of his title were, principally, the fossils recorded by contemporary palaeontology, a record which Chambers summarised with erratic accuracy in the central part of the book. His proposal for a natural explanation of the 'creation' of species is perhaps nearest to Geoffroy's in its use of embryonic monstrosities, its emphasis on the direct influence of the environment and its acceptance of trans-specific 'jumps'; though whether Chambers knew of Geoffroy's work at first hand or not is uncertain, and it seems rather unlikely since most of his direct sources were in English. But in any case his theory also involved an astonishing hotchpotch of credulous speculations on, for example, the origin of life by "a chemico-electric operation" (a fashonable scientific catch-phrase of the time), and on the spontaneous generation of mites and the spontaneous mutation of rye into wheat. It is therefore understandable that the general reaction of the scientific community should have been the same kind of irritated exasperation that has been evoked by many works of 'pop science' in our own day. When Chambers suggested, for example, that the bony armour of the Old Red Sandstone fish was much like the external skeletons of arthropods, and that the first fish might therefore have been evolved from arthropods; or when he made liberal use of the alleged three-fold parallel between the embryonic

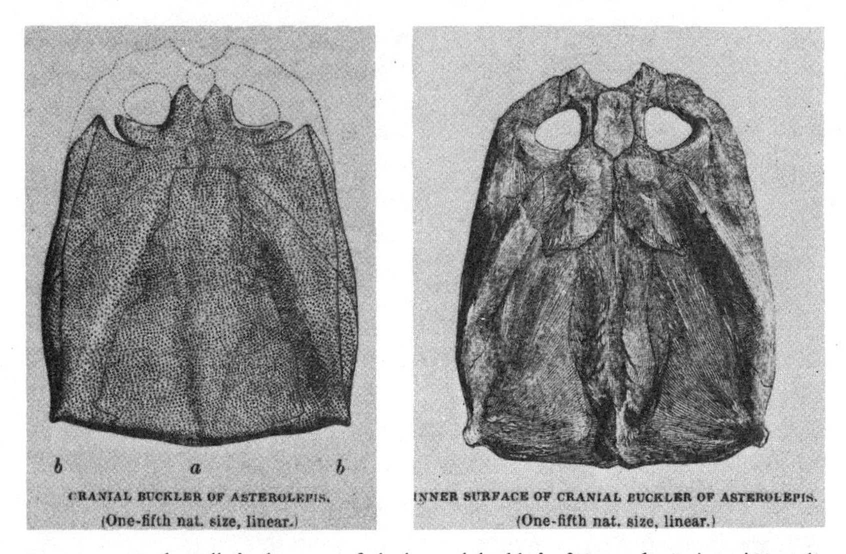

Fig. 4.8. Hugh Miller's drawings of the 'cranial buckler' of Asterolepis *(1847)*[44]. *The structural complexity of this ancient fish from the Old Red Sandstone was used by Miller to argue against Robert Chambers's evolutionary theory, according to which the earliest vertebrates should have been relatively simple and 'primitive'.*

development of the individual from simple to complex, the supposed scale of organisms from 'lower' to 'higher', and the 'progress' of the fossil record from the earliest forms of life to the most recent: on reading such arguments most scientists naturally felt that these were just the kind of irresponsible speculations that careful comparative anatomy and embryology had exploded once and for all.

Yet the almost violent reaction to Chambers's book suggests that he had done more than wound the professional pride of the scientists. That Chambers represented a more fundamental threat is suggested by the appearance soon afterwards of a comparable popularising work designed to refute the *Vestiges*. Coming from the very same social milieu, the *Footprints of the Creator* (1849) indicates that behind the vehemence of the scientific critics of the *Vestiges* lay the fear that Chambers's naturalistic explanation of 'creation' threatened the status of Man himself. Hugh Miller (1802–1856), another Scottish journalist but one with the advantage of first-hand knowledge of palaeontology, subtitled his book *The Asterolepis of Stromness*[44]; and he used his own research on the Old Red Sandstone fish to undermine Chambers's theory at one its weakest points, by showing that these earliest vertebrates were quite as complex and highly adapted as any later fish, and not at all the primitive organisms that Chambers assumed (Fig. 4.8). However Miller argued quite explicitly that beyond the scientific arguments what was really at stake was the dignity of Man as a responsible moral agent: if Man had come into being by the chance process that Chambers implied, he could not be held morally responsible for his actions, and the whole fabric of society was thereby threatened. It was this, lying behind even the threat of such theories to the sense of the wise design of the organic world, that imbued the evolutionary debate with depth and seriousness.

X

It does not follow, however, that in rejecting Chambers's evolutionary theory the palaeontologists of the 1840s were left without any rationally satisfying explanatory framework for their work. On the contrary, although Darwin later suggested that the only alternative to his evolutionary theory was a naive creationism, in fact there was another explanation available, with intellectual credentials quite as high as Darwin's, and with considerably more credibility to the mind of the time. This alternative was well developed by one who started as Darwin's collaborator but who later became one of his most implacable opponents—the anatomist and palaeontologist Richard Owen (1804–1892).

Owen was above all a comparative anatomist, not only by the circumstances of his professional career—he was at this time Hunterian Professor at the Royal College of Surgeons in London—but more deeply by the whole direction of his scientific interests. The cognitive passion behind his extraordinary scientific energy was the desire to understand the forms and functions of animals, indeed to get behind specific forms and particular functions to an understanding of the fundamental nature of the animal kingdom. As one possible level of understanding, Owen was not at all hostile to the general idea of evolution. Unlike most of his scientific friends he did not join the general chorus of condemnation of the *Vestiges:* on the contrary, he approved its general intention, and said that he himself was then toying with no less than six alternative hypotheses for the production of new species. However in his view the means by which new forms of life came into existence would hardly touch the deeper problems of the diversity of the animal kingdom, as embodied in their natural history. For Owen, as for most of his contemporaries, the concept of 'Natural History' was still meaningful and unified: it was not yet demoted to the derogatory status it now holds for many scientists. 'Natural History' was still, as it had been for Linnaeus and Buffon in the 18th century, the systematic ordering of the whole range of diverse natural entities. That some of these entities were what we would now call organisms, and others inanimate, was quite irrelevant: the diversity of the Mineral Kingdom needed to be ordered in just the same sense as that of the Vegetable and Animal Kingdoms. Classification was not a means to an end, a clue to evolutionary relationships, for example: it was itself an end, the end of knowing the true order of Nature.

This mode of knowledge is not easy for us to recapture from our post-Darwin standpoint. To us the diversity of animals and plants seems an entirely different problem from the diversity of minerals, because we believe the first has been produced over time by entirely different means. Yet this way of knowing nature that was termed 'Natural History' was not old-fashioned in Owen's time, even though it had had a long history. Owen himself was not a kind of living fossil epistemologically: his view of nature was that of most of his contemporaries. The practical effect of this viewpoint was not only to remove the problem of the origin of species from the central position it occupied in the mind of Darwin, and to make it seem less pressing; it also focussed Owen's attention on the kinds of order and the kinds of diversity that natural entities in general, and animals in particular, displayed. It pre-disposed him moreover to see these problems in static rather than dynamic terms, even though as a highly competent palaeontologist he was well aware of the time dimension revealed by geological research. He was more concerned with the pattern of organic diversity than with the way in which it had come into being; more interested in the nature of functional adaptation than with its origin.

Owen believed that there were two distinct components to the problem of organic order and diversity, and his integration of the two was felt by him and by most of his contemporaries to be his greatest and most lasting achievement. The first component was the adaptation and design manifest in the structure and function of the individual organism. Owen's strong sense of designful functional adaptation derives directly and very clearly from his hero Cuvier: and the heuristic value

Fig. 4.9. A contemporary sketch (1854) of some of the reconstructed fossil animals set up by Richard Owen in the grounds of the Crystal Palace[45].

of the Cuvierian tradition comes out most clearly and spectacularly in Owen's work, as in Cuvier's before him, in the reconstruction of *fossil* vertebrates.

Ironically, Owen's best known fossil reconstruction was one in which his confidence in the Cuvierian method overreached itself. When the 1851 Exhibition building was re-erected as the Crystal Palace, Owen conceived, and Benjamin Hawkins executed, a series of life-size reconstructions of "the inhabitants of the ancient world" for exhibition in the grounds (Fig. 4.9). As an effective publicity stunt, Hawkins invited Owen and twenty other guests to a dinner to be held inside his reconstruction of the *Iguanodon*[45]. Owen had had to work from very incomplete material in this case, and he had reconstructed the *Iguanodon* as a lumbering quadruped instead of the huge but rather elegant bipedal reptile that it turned out to be, when complete skeletons were discovered later in the century. But generally Owen's predictions were sound. Perhaps the most spectacular example was his prediction of the character of the Moas of New Zealand. On the basis of a single short fragment of femur, Owen in 1838 predicted that it would be found to belong to a very large flightless bird. The editors at the Zoological Society understandably had doubts about publishing such an apparently hazardous and weakly-founded idea; but Owen was brilliantly vindicated five years later by the arrival of relatively complete skeletons[46]. The same Cuvierian principles were used by Owen in his work, at the same early period, on the extinct mammals that Darwin had brought back from South America. This earned him the Geological Society's highest award, the Wollaston Medal, in 1838; and the Royal Society likewise gave him their Royal Medal in 1846 for his brilliant interpretation of the functional anatomy of the belemnites. All this work, and much more, which has largely endured to the present, was achieved during this period under the dominant influence of the belief that all these diverse organisms, by whatever *means* they had been formed, were "specimens of divine mechanism". Organisms were not less mechanistic for being manifestations of the Creative Power.

This, then, was one component in the problem of organic form, as it was interpreted by Owen and most of his contemporaries: namely, the perception of designful adaptation in the structure of every organism. But how was this related to the diversity of different organisms? Anatomically, Cuvier's four major groups—Vertebrates, Molluscs, Articulates, and the lower invertebrates—remained for Owen as sharply distinct from each other as they had been to Cuvier; but Geoffroy's principle of 'unity of composition' seemed increasingly to be valid *within* each of these groups. Within the vertebrates, for example, the equivalent bones *could* be recognised in the foreleg of a lizard, the wing of a bird, the foreleg of a tiger, the flipper of a seal and the arm of a man. Such comparisons, which Owen termed *homologies*, had of course been made since Antiquity; but erroneous applications of the principle had brought it into disrepute. Owen in his comparative anatomy pursued the principle with greater thoroughness and critical care that it had ever previously been given; and he was able to identify the homologues of many of the bones of the vertebrate skeleton more successfully than ever before.

The determination of homologies was not, however, a simple empirical exercise, any more than the functional interpretation of the same structures. It too was underlain by a deep metaphysical belief. Owen was convinced that homologies could be traced in animals of different classes, because he believed that all the animals within each major group were variations on a single theme, modifications of a single Ideal Type. As early as 1841 Owen was beginning to talk about the possibility of discovering the "Ideal Archetype" on which all the diversity of the vertebrates was based. This was important enough for the

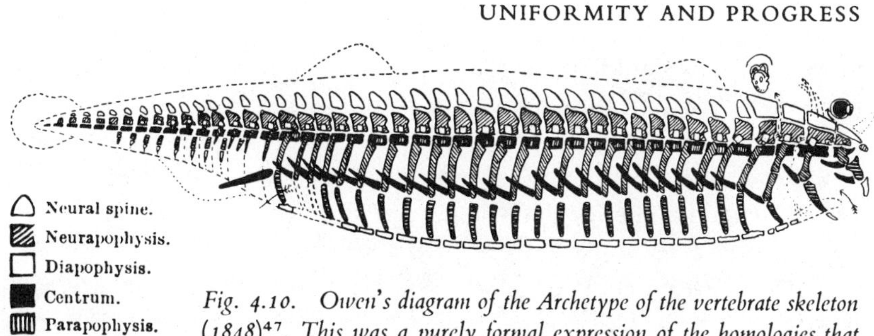

Fig. 4.10. Owen's diagram of the Archetype of the vertebrate skeleton (1848)[47]*. This was a purely formal expression of the homologies that could be traced between the skeletons of actual vertebrates, living and fossil. It was* not *intended to depict a viable common ancestral form, but Owen did believe that the fossil record revealed the embodiment of this archetype in progressively 'higher' and more diverse forms, culminating in Man. (From the diagram as re-drawn by Russell.)*

British Association to commission him to write a Report on the subject; and in 1848 he published his most important theoretical treatise *On the Archetype and Homologies of the Vertebrate Skeleton.* The previous year a translation had been published of the idealistic natural philosophy of Lorenz Oken, the founder of the German organisation on which the British Association had been modelled, and a friend and disciple of Schelling and other romantic philosophers. Owen felt at once that Oken, years before, had intuitively recognised and sketched out the principles which he, Owen, had now established by more rigorous and detailed comparative anatomy. There *was* a 'unity of composition' within the vertebrates: the bones of a typical mammal, for example, could all be recognised and identified in the skeletons of the other classes, and all the diversity of the vertebrates could be derived from a single Ideal Type, the Archetype (Fig. 4.10). The Archetype was not an animal that ever had existed or could exist: it was the Platonic structural Idea, of which all actual vertebrates were functionally diverse embodiments[47].

Owen's discovery, by patient comparative anatomy, of what he believed to be the true Archetype of the highest group of animals, the group to which Man himself belonged, was in his own eyes his supreme achievement. Cuvierian functional anatomy revealed the designful adaptiveness of animals, but it failed to account for the underlying homologies of structure that could not be attributed to functional similarity. Cuvierian principles, for example, could explain the marvellous adaptation of the wings of birds and bats as mechanisms for flight; but they could not explain how or why those wings had been formed by the different modifications of the same skeletal elements. But seen in the light of the Archetype, the diversely adapted fore-limbs of all the vertebrates became manifestations of the same single pre-existent Platonic Idea.

It may seem a long way from such abstract high-flown speculations to the mundane details of palaeontological research. But in Owen's mind the distance was not so great, for he saw the diversity of living organisms as the result, *through time*, of the gradual embodiment of underlying pre-existent Ideas; and it was palaeontology that gave concrete evidence of that process. The vertebrate Archetype was not the common ancestor of the vertebrates, but a divine Idea which had been embodied in successive, marvellously diverse and intricately adapted forms of real life. Throughout, there was a plan or programme leading, however erratically, towards its culmination in Man; and this plan was the true plan of nature.

Owen's viewpoint is beautifully summarised in the concluding passage of his lecture *On the Nature of Limbs*, given in 1849 just ten years before Darwin published the *Origin of Species:*

> The recognition of an ideal Exemplar for the Vertebrated Animals proves that the knowledge of such a being as Man must have existed before Man appeared. For the Divine mind which planned the Archetype also foreknew all its modifications. The Archetypal idea was manifested in the flesh, under divers modifications, long prior to the existence of those animal species that actually [that is, now] exemplify it. To what natural or secondary causes the orderly succession and progression of such organic phenomena may have been committed, we are as yet ignorant. But if, without derogation to the Divine Power, we may conceive the existence of such ministers and personify them by the term *Nature*, we learn from the past history of our globe, that she has advanced with slow and stately steps, guided by the archetypal light amidst the wreck of worlds, from the first embodiment of the vertebrate idea, under its old ichthyic vestment [that is, as fish], until it became arranged in the glorious garb of the human form.[48]

To this passage, with its expression of progressive incarnation within a strange fusion of Christianity and Platonism, our twentieth century reaction may well be, 'C'est magnifique, mais ce n'est pas la science'. But Darwin too ended his book with a piece of purple prose, remarking

Formations	circa 1790	circa 1840		Modern	
Alluvium	Post-diluvial	Alluvium	CAINOZOIC Tertiary	Holocene	CAINOZOIC Tertiary
Glacial deposits	Diluvial	Newer Pliocene		Pleistocene	
Sicilian strata		Older Pliocene		Pliocene	
				Miocene	
Subappenine strata	TERTIARY	Miocene		Oligocene	
				Eocene	
Parisian strata		Eocene		Palaeocene	
Chalk			MESOZOIC		MESOZOIC
		Cretaceous	Secondary	Cretaceous	
Oolites					
Lias		Jurassic		Jurassic	
Muschelkalk New Red Sandstone	SECONDARY				
		Triassic		Triassic	
Kupferschiefe		Permian		Permian	
Coal Measures			PALAEOZOIC		PALAEOZOIC
		Carboniferous		Carboniferous	
Carboniferous (Mountain) Limestone					
Old Red Sandstone		Devonian		Devonian	
Wenlock Limestone					
				Silurian	
'Grauwacke'	TRANSITION	Silurian	(Protozoic)		
				Ordovician	
		Primordial (Cambrian)		Cambrian	
Longmynd strata				PRE-CAMBRIAN	
Scandinavian schists etc.	PRIMARY	AZOIC (Primary)			

Fig. 4.11. Chart to illustrate the division of the geological time-scale about 1840, with modern equivalents. Some characteristic formations mentioned in the text are shown in the left-hand column. The major divisions recognised about 1790 are given in the second column; but it should be noted (a) that much more detailed divisions were being applied at this date to particular regions, although correlation between them was uncertain, and (b) that the categories 'Transition' and 'Primary' have no exact equivalents in later schemes, because they included slightly or highly metamorphosed rocks of many ages (though predominantly of the ages shown).

that "there is grandeur in this view of Nature". There was indeed grandeur in Darwin's view, but so too is there in Owen's. Certainly it is impossible to understand the generally critical reception of the *Origin of Species*, and Owen's opposition in particular, without recognising the intellectual strength of the alternative view of nature that Owen's work epitomised. In its relative lack of interest in the causal mechanism by which organic diversity was produced, and in its reliance on an idealistic metaphysics that later biology rejected, this alternative may seem to us remote and implausible; but at the time it was both persuasive and attractive. Above all, it interpreted the evidence of fossils and of living organisms in terms of a Nature that was harmonious, integrated and designful, and that had developed over the aeons of geological time according to an intelligible and meaningful Plan.

REFERENCES

1. Mrs. Lyell (ed.), *Life Letters and Journals of Sir Charles Lyell, Bart.*, London, 1881: vol. 1, pp. 233–4.
2. H. B. Woodward, *The History of the Geological Society of London*, London, 1907; M. J. S. Rudwick, 'The foundation of the Geological Society of London: its scheme for co-operative research and its struggle for independence', *British Journal for the History of Science*, vol. 1, pp. 325–355 (1963).
3. See Roy Porter, 'The Industrial Revolution and the rise of the science of geology', *in* M. Teich and R. Young (eds.), *Changing Perspectives in the History of Science*, London, 1973, pp. 320–343.
4. See Chap. 3, note 16.
5. Charles Lyell and Roderick Impey Murchison, 'On the excavation of valleys, as illustrated by the volcanic rocks of Central France', *Edinburgh new philosophical Journal*, vol. 12, pp. 15–48 (1829).
6. C. A. Basset, *Explication de Playfair sur la Théorie de la Terre par Hutton, et Examen comparatif des systêmes géologiques fondés sur le feu et sur l'eau, par M. Murray; en réponse à l'Explication de Playfair*, Paris, 1815.
7. Karl Ernst Adolf Hoff, *Geschichte der durch Überlieferung nachgewiesenen natürlichen Veränderungen der Erdoberfläche*, 3 vols, Gotha, 1822–34; see also Zittel, *Geschichte*, p. 285
8. G. Poulett Scrope, *Considerations on Volcanos, the probable causes of their phenomena, the laws which determine their march, the disposition of their products, and their connexion with the present state and past history of the globe; leading to the establishment of a new Theory of the Earth*, London 1825; *Memoir on the Geology of Central France, including the volcanic formations of Auvergne, the Velay and the Vivarais*, London, 1827.

9. See Chap. 3, note 49; also Fleming's 'Remarks illustrative of the influence of Society on the distribution of British animals', *Edinburgh philosophical Journal*, vol. 11, pp. 287–305 (1824).

10. [Charles Lyell,] 'Memoir on the Geology of Central France... by G. P. Scrope...', *Quarterly Review*, vol. 36, pp. 437–483 (1827).

11. Lyell and Murchison, 'Excavation of Valleys' (note 5); Croizet et Jobert, *Recherches sur les ossemens fossiles du Département du Puy-de-Dôme*, Paris, 1826–8.

12. See M. J. S. Rudwick, 'Lyell on Etna, and the antiquity of the Earth', *in* Schneer, *Toward a History of Geology*, pp. 288–304.

13. Lyell, Life, *Letters and Journals*, vol. 1, p. 234.

14. Charles Lyell, *Principles of Geology, being an attempt to explain the former changes of the earth's surface, by reference to causes now in operation*, 3 vols, London, 1830–3 (facsimile reprint, New York, 1970); see also Martin J. S. Rudwick, 'The strategy of Lyell's *Principles of Geology*', *Isis*, vol. 61, pp. 4–33 (1970).

15. See Walter F. Cannon, 'The impact of uniformitarianism. Two letters from John Herschel to Charles Lyell, 1836–1837', *Proceedings of the American Philosophical Society*, vol. 105, pp. 301–314 (1961).

16. Lyell, *Life, Letters and Journals*, vol. 1, p. 251.

17. W. D. Conybeare, 'Report on the Progress, Actual State, and Ulterior Prospects of Geological Science', *Report of the British Association for Advancement of Science*, vol. for 1831–2, pp. 365–414 (1833).

18. [William Whewell], '*Principles of Geology* . . . By Charles Lyell . . . Vol. I . . . ', *British Critic*, vol. 9, pp. 180–206 (1831); '*Principles of Geology* . . . By Charles Lyell . . . Vol. II . . . ', *Quarterly Reveiw*, vol. 47, pp. 103–132 (1832). See also Walter Cannon, 'The problem of miracles in the 1830s', *Victorian Studies*, vol. 4, pp. 4–32 (1960).

19. Conybeare, 'Report on . . . Geological Science' (note 17).

20. Adam Sedgwick, 'Address to the Geological Society . . . Feb. 18, 1831', *Proceedings of the Geological Society of London*, vol. 1, pp. 281–316 (1831).

21. J. F. W. Herschel, 'On the astronomical causes which may influence geological phenomena', *Transactions of the Geological Society of London*, 2nd series, vol. 3, pp. 293–299 (1832). On Herschel's prestige, see Walter F. Cannon, 'John Herschel and the idea of science', *Journal of the History of Ideas*, vol. 22, pp. 215–239 (1961).

22. [Scrope], '*Principles of Geology* . . . By Charles Lyell . . . Vol. I', *Quarterly Review*, vol. 43, pp. 411–469 (1830); '*Principles of Geology* . . . By Charles Lyell . . . 3rd Edition', *ibid.*, vol. 53, pp. 406–448 (1835).

23. Darwin, *Extracts from Letters addressed to Professor Henslow*, Cambridge, 1835; 'Observations of proofs of recent elevation on the coast of Chili . . . ', *Proceedings of the Geological Society of London*, vol. 2, pp. 446–449 (1837).

24. Charles Darwin, 'On certain areas of elevation and subsidence in the Pacific and Indian oceans, as deduced from the study of coral formations', *Proceedings of the Geological Society of London*, vol. 2, pp. 552–554 (1837); *The Structure and Distribution of Coral Reefs* . . . , London, 1842.

25. Deshayes, 'Tableau comparatif des espèces de coquilles vivantes avec les espèces de coquilles fossiles des terrains tertiaires de l'Europe, et des espèces de fossiles de ces terrains entr'eux', *Bulletin de la Societé géologique de France*, vol. 1, pp. 185–187 (1831); Heinr. G. Bronn, *Italiens Tertiär-Gebilde und deren organische Einschlüsse. Vier Abhandlungen*, Heidelberg, 1831.

26. Adam Sedgwick, 'Address delivered at the Anniversary Meeting of the Geological Society of London, on the 19th February, 1830', *Proceedings of the geological Society of London*, vol. 1, pp. 187–212 (1830).

27. R. I. Murchison, 'On the sedimentary deposits which occupy the western parts of Shropshire and Herefordshire, and are prolonged from N.E. to S.W., through Radnor, Brecknock and Caermarthenshires, with descriptions of the accompanying rocks of intrusive or igneous characters', *Proceedings of the geological Society of London*, vol. 1, pp. 470–477 (1833); 'On the Silurian system of rocks', *Philosophical Magazine and Journal of Science*, vol. 7, pp. 46–52 (1835).

28. Roderick Impey Murchison, *The Silurian System, founded on geological researches in the counties of Salop, Hereford, Radnor, Montgomery, Caermarthen, Brecon, Pembroke, Monmouth, Gloucester, Worcester and Stafford; with descriptions of the coal-fields and overlying formations*, London, 1839.

29. Leonard G. Wilson, 'The emergence of geology as a science in the United States', *Cahiers d'Histoire mondiale*, vol. 10, pp. 416–437 (1967).

30. M. J. S. Rudwick, 'The Devonian System 1834–1840. A study in scientific controversy', *Actes du XIIme Congrès international d'Histoire des Sciences*, vol. 7, pp. 39–43 (1971).

31. R. I. Murchison, Édouard de Verneuil and Alexander von Keyserling, *The Geology of Russia in Europe and the Ural Mountains*, vol. 1, London, 1845.

32. Sedgwick and R. I. Murchison, 'On the *Silurian* and *Cambrian* Systems, exhibiting the order in which the older Sedimentary Strata succeed each other in England and Wales', *Report of the British Association for the Advancement of Science*, vol. for 1835, *Transactions of the Sections*, pp. 59–61 (1836).

33. Murchison *et al.*, *Geology of Russia*, Introduction.

34. J. Barrande, *Notice préliminaire sur le Système Silurien et les Trilobites de Bohême*, Leipzig, 1846.

35. Buckland, 'Address to the Geological Society . . . 21st of February, 1840', *Proceedings of the geological Society of London*, vol. 3, pp. 210–267 (1840).

36. Louis Agassiz, 'On a new classification of fishes, and on the geological distribution of fossil fishes', *Proceedings of the geological Society of London*, vol. 2, pp. 99–102 (1834); *Recherches sur les Poissons fossiles* . . . , Neuchatel, 1833–1843. See also Edward Lurie, *Louis Agassiz: a life in science*, Chicago, 1960.

37. John Phillips, *Figures and Descriptions of the Palæozoic Fossils of Cornwall, Devon and West Somerset: observed in the course of the Ordnance Geological Survey of that District*, London, 1841; see 'Notices and Inferences', pp. 155–182.

38. Charles Lyell, *Elements of Geology*, London, 1838 (and later editions). Gideon Algernon Mantell, *The Wonders of Geology; or, A Familiar Exposition of Geological Phenomena*, London, 1838; *Medals of Creation; or, First Lessons in Geology, and in the Study of Organic Remains*, London, 1844; *Thoughts on a Pebble, or, A First Lesson in Geology*, London, 1849.

39. Murchison, *et al.*, *Geology of Russia*, vol. 1, see Conclusion.

40. See note 15.

41. William Buckland, *Geology and Mineralogy considered with reference to natural theology*, 2 vols, London, 1836.

42. R. I. Murchison, Edouard de Verneuil et Alexandre de Keyserling, *Géologie de la Russie d'Europe et des Montagnes d'Oural*, [Vol. 2], Paris, 1845: see Avant-Propos.

43. [Robert Chambers], *Vestiges of the natural history of creation*, London, 1844 (and many later editions: facsimile reprint of first edition, Leicester, 1970); *Explanations: a Sequel to 'Vestiges of the Natural History of Creation'*, London, 1845.

44. Hugh Miller, *Foot-Prints of the Creator: or, the Asterolepis of Stromness*, Edinburgh, 1847; see also *The Old Red Sandstone; or New Walks in an Old Field*, Edinburgh, 1841; W. M. Mackenzie, *Hugh Miller. A Critical Study*, London, 1905.

45. Richard [S.] Owen, *The Life of Richard Owen*. 2 vols, London, 1894; Richard Owen, *Geology and Inhabitants of the Ancient World*, London, 1854 (Crystal Palace Guidebooks).

46. Owen, 'Notice of a fragment of the femur of a gigantic bird of New Zealand', *Transactions of the Zoological Society of London*, vol. 3, pp. 29–32 (1842); 'On the Dinornis, an extinct genus of tridactyle struthious birds, with descriptions of portions of the skeleton of five species which formerly existed in New Zealand (Part I)', *ibid.*, vol. 3, pp. 235–275 (1844). Owen's inference was less straightforward than he made it appear: see C. F. A. Pantin, *Science and Education*, Cardiff, 1963, pp. 19–26.

47. Richard Owen, *On the Archetype and Homologies of the Vertebrate Skeleton*, London, 1848; [article] 'Oken', *Encyclopaedia Britannica*, 8th edition, vol. 16, pp. 498–503 (1858); see also Russell, *Form and Function*, ch. 8, and Roy M. MacLeod, 'Evolutionism and Richard Owen, 1830–1868: an episode in Darwin's century', *Isis*, vol. 56, pp. 259–280 (1965).

48. Richard Owen, *On the Nature of Limbs*, London, 1849.

49. Richard Owen, *A History of British fossil Mammals, and Birds*, London, 1846.

Chapter Five

Life's Ancestry

I

ON 2 February 1857, at a public meeting of the Academy of Sciences in Paris, the award of a Grand Prix for 'Physical Sciences', a gold medal to the value of 3000 francs, was announced. The prize, offered seven years before for an essay on the 'laws of development' of the organic world, had been won by a candidate whose work bore the motto "To be taught by Nature" (*Natura doceri*). The adjudicating committee, which included Élie de Beaumont, Geoffroy's son Isidore, and the younger Brongniart, would have had little difficulty in piercing the formal anonymity of the motto, and in guessing that the author was in fact Heinrich-Georg Bronn, the Professor of Natural History at Heidelberg[1]. Probably no other palaeontologist at this period had the breadth of experience to be able to synthesise in so masterly a manner the current knowledge of the distribution of organisms in geological time, or to distil from this information so valuable a set of generalisations. By the canons of an earlier historiography Bronn was on the 'losing side', for he doubted the reality of trans-specific *evolution* in the modern sense of the word, and proposed no satisfactory causal explanation of the trends he observed in the history of life. But without the kind of synthesis that Bronn attempted it is doubtful indeed whether palaeontologists would have been able, within the space of a few years, to accept from Darwin the theory that evolution *had* occurred, even though Bronn's synthesis may also have contributed to their continued—and in many respects justified—scepticism about the adequacy of the mechanism Darwin proposed.

The terms in which the prize-question was set in 1850 are an illuminating commentary on the state of palaeontological debate at that time. The essay was "to study the laws of the distribution of fossil organisms in the different sedimentary strata according to the order of their superposition; to discuss the question of their successive or simultaneous appearance or disappearance; to examine the nature of the relations between the present and former states of the organic world"[2]. The committee emphasised the new urgency of these problems and the scientific importance of finding satisfactory answers to them—answers which could only be discovered by reference to the fossil record.

The new situation was due indirectly to the rapid growth in the sheer volume of factual information about strata and their fossils, for it was this that had highlighted the major problems involved. For example, the description of new faunas and floras, and the recognition of new episodes of tectonic activity in the history of the Earth's crust, had been modifying the concept of successive 'revolutions' to the point of vagueness[3]. If, following Élie de Beaumont (the assessor of the committee), tectonic revolutions were regarded as episodes of relatively localised mountain elevation, how could they have caused the sudden destruction of whole faunas and floras—especially marine faunas—on a global scale? *A fortiori*, if one followed Lyell's actualistic principle and allowed no 'revolutions' more intense than those witnessed in human history, what could have caused these episodes of apparent mass extinction? Or were they more apparent than real? Just how sharp were the faunal and floral breaks between successive formations: how many species were really confined to a single formation, and how many had ranges that straddled the supposed revolutions? Even for those confined to a single formation, was this a measure of their real range in time, or a reflection of a fragmentary geological record: had such species originated just before the deposition of that formation and become extinct just after deposition ended, or had they had a longer existence that was only partially represented by the strata in which they were found as fossils?

The stratigraphical ranges of fossil species were becoming difficult to determine with precision, because the notion of the species itself had become problematical in practice. Well-known and established fossil species were becoming subject to taxonomic 'splitting' as they were studied in further detail; but conversely the increasing specialisation among palaeontologists could easily lead to a single species being given different names in different formations or systems of strata. In any case, where a new species appeared to have replaced a similar one found in earlier strata, was this due to a new 'creation' (whatever that meant precisely) or to some kind of transmutation or (in the modern sense of the word) evolution?

Finally, even if some kind of transmutation were acceptable between similar species, could such an hypothesis be extrapolated into a full-blown concept of 'development' from the simplest 'monads' to the most complex mammals, as Lamarck had supposed? What empirical support was there for the idea of 'progressive development' in the course of time? Was it restricted to the progressive appearance of 'higher' forms *within* a given major group? If 'development' was a reality, had it been correlated with demonstrable changes in the inorganic environment, or did it show signs of being independent of the environment and somehow inherent in the organisms themselves?

In view of the inherent importance of factual data in any satisfactory treatment of these problems, Bronn's successful candidature is hardly surprising. Indeed, his outstanding qualifications for the task, and the fact that he won the prize, illustrate neatly the professionalisation of palaeontology during the middle decades of the century. The heyday of the gifted gentleman-naturalists was over (Darwin, here as in so many other respects, is the exception that proves the rule). Even Murchison, the tireless collector of honours and insignia from all the courts of Europe, had entered what was to become the scientific civil service, succeeding De la Beche in the administrative post of director of the British Geological Survey. Even in Britain, but far more on the Continent and in the booming United States of America, the bulk of palaeontological research was passing more and more into the hands of those who were paid to do it, whether as employees of state geological surveys or museums, or as university teachers and researchers. Bronn's career is typical of this new professionalised class of scientist. He had passed directly from being a student to a junior teaching post at Heidelberg, and became a professor there while still in his twenties. In 1830 he had become an editor of the famous *Neues Jahrbuch*, one of the most important journals for the publication of geological and palaeontological research; and he remained an editor until his death more than thirty years later. It was a strategic position that not only gave him great influence within the profession, but also enabled him to keep pace with the unprecedented 'information explosion' in the science. This he exploited to the full throughout his long academic career, publishing a prodigious output of works that were at once rich in compilative detail and suggestive in their theoretical exploration[4]. His prize essay was the culmination of this series of palaeontological syntheses.

The fact that Bronn, a German, was awarded so high a prize by the French Academy of Sciences is also a significant illustration of the internationalisation of the scientific community during this period. Paradoxically, this coincided with the rampant growth of 'national consciousness' and the political evolution of the modern nation-state. The movement for the unification of the German peoples, for example, had been reflected in Oken's consciously political Assembly of German Scientists and Doctors (founded 1822), which was the prototype of the British Association. But as early as 1840 Murchison, for example, was airing the suggestion that such national bodies should occasionally replace their separate annual meetings with a single international conference of scientists; and in many fields of science international scientific projects were already well under way. International 'commuting' by distinguished scientists had become almost as commonplace (though not as rapid) as in our own day, and the award of high scientific honours by one country to the citizens of another was established practice long before Bronn's prize was announced. Nevertheless, Bronn's victory serves conveniently to underline the full internationalisation of the scientific community by the middle of the century—a process that was, for obvious reasons, particularly important in a science such a palaeontology.

II

Even in some of his earliest work, already mentioned for its parallel with Lyell's, Bronn had attempted to apply numerical techniques to the problem of assessing degrees of affinity between the faunas of different formations; and all his later compilative works illustrate the same belief that organic change is always and everywhere an essentially gradual process, taking place by piecemeal extinction of old species and production of new. However, he had never made Lyell's mistake of assuming that the 'uniformity' of gradual change necessarily entailed that of a steady-state picture of the history of life. Gradual change might well have been combined with an overall directional development: it was for the facts themselves to determine whether or not this was so. Bronn's self-conscious empiricism—admittedly a commonplace among scientists at this period—was deliberately epitomised in his choice of a motto for his prize essay: "To be taught by Nature". But it was not a barren empiricism that eschewed all theorising. On the contrary, Bronn marshalled the evidence that Nature had taught him, in order to derive generalised 'laws' of Nature's working—as indeed the terms of the prize required.

After introducing the factual evidence in the form of massive tables, Bronn plunged at once into the heart of all the theoretical problems: the nature of the "creative force" (*Schöpfungs-Kraft*) responsible for the production of new species. Bronn's use of this phrase is extremely important, for it indicates the terms in which the problem of the origin of species was commonly being discussed at the time that Darwin was preparing to publish his work on the subject. Whether from ignorance of Continental science, or in order to strengthen the persuasiveness of his case, Darwin presented his theory with only a straw man to oppose it: *either* slow trans-specific evolution by means of 'natural selection', *or* direct divine creation of new species from the inorganic dust of the Earth. So successful was Darwin's statement of his case along these lines that later scientists and historians, at least in English-speaking countries, have often taken his essentially polemical argument at its face value. But in fact the debate was never so sharply polarised. When scientists such as Richard Owen protested after 1859 that they were quite prepared to accept evolution of some kind but not Darwin's mechanism for it, they were not hastily retreating to a new untenable position, but merely restating a view they had held before Darwin ever introduced the idea of natural selection. It is a mistake, therefore, to leap to the conclusion that Bronn's word 'creative' implies that he believed in the 'special creation' of every species. On the contrary, he explicitly denied this, taking (like Darwin!) the common nineteenth-century view that a Creator achieving his creative effects through the use of 'secondary' agencies would be more sublime than One who was obliged to interfere continuously or periodically with his creation. The 'creative force' was creative in the sense that its effects were innovative; but it was most emphatically a *natural* agency.

The real difficulties began, of course, when one tried to define what *kind* of natural agency it could be. Bronn was well aware of Lamarck's theory of transmutation, of Geoffroy's modification of it, and of its continued support from a minority of his own scientific contemporaries, such as the botanist Franz Unger[5] (he did not mention, and may not have known, Chambers' pop version). However, all such theories seemed to Bronn to founder on the rocks of hard facts: there was simply no actualistic *evidence* that intra-specific variation ever led to more than temporarily distinct races, or that even the simplest forms of life were ever generated 'spontaneously' from non-living materials; and paleontology conspicuously failed to provide any fossil evidence either.

With transmutation scientifically unacceptable, Bronn turned in a different direction, and tried to advance some way towards a solution by utilising the methodological tradition of Newtonian physics. This is where the second word in Bronn's phrase—"force"—needs correct interpretation: the word's associations are physical, not 'vitalistic'. Bronn used explicitly an analogy between the creative force and the physical forces of gravitation and chemical affinity. We may remain ignorant of the ultimate nature of all such forces, he argued, but we can still try to understand them by studying their effects. Even at its earliest manifestation, Bronn maintained, the creative force's "primordial productions" were diverse and in their own way 'perfect': in other words the 'Primordial' or early 'Silurian' species were not imperfect rudiments of organisms but as clearly adapted to their environment as any later species. Subsequent geological history had been marked by the continued effects of the creative force, apparently working according to a kind of rational plan. This plan, or pattern in the history of life, was characterised by a continued replacement of extinct species by more diverse and also 'higher' forms of life, while maintaining at all times a well-balanced ecological assemblage.

However, any such formulation begged some very large questions about the concepts used, and Bronn therefore tried to clarify the two major variables in his interpretation. The first was the concept of 'comparative degrees of perfection': what precise meaning could be attached to the semi-intuitive, semi-traditional biological notion of a hierarchy of organic forms? What was meant by 'lower' and 'higher', 'simple' and 'complex', whether in relation to different groups of organisms, different forms of organs or of whole organisms, or different stages in the growth of an individual organism[6]; and how was the notion of hierarchy related to the notion of adaptation to the environment? Without some clarification here, it would be impossible to evaluate the fossil record as possible evidence of the 'progressive development' of life. The second fundamental variable was geological: what could 'development' mean in relation to the external environment of organisms? How had the Earth's surface changed in the course of geological time as a potential habitat for living organisms?

Bronn tried to show that the concrete evidence of paleontology illustrated a series of ten "secondary laws" or generalisations about the history of life, all of which involved one or both of these fundamental variables. Having suggested, on the basis of the best evidence available, that plants and animals first appeared at the same (Primordial or earliest Silurian) period of Earth-history, as might be anticipated from their ecological inter-dependence, he then correlated the subsequent fossil record with the geological evidence for a gradually cooling Earth, showing how the successive faunas and floras indicated a gradual lowering of temperature and increasing climatic diversification. Arguing for the reality of the species as an objective natural unit, he then showed that the production and extinction of species had occurred continuously and with only minor oscillations in tempo, so that the organic world had come quite gradually to approximate more and more to its present state: in other words, there had been no periods of massive extinction or of wholesale production of new species. In part, this development had been one of increasing taxonomic diversity, which could be correlated with the progressive physical diversification of the Earth's surface and, indirectly, with the consequent increasing complexity of the ecological relations between organisms. However the development of the organic world also showed a distinct "terripetal" trend towards an increasing colonisation of terrestrial habitats, and furthermore a true "progressive" trend towards more complex forms of life within each of the major groups.

From all these generalisations, grounded in detailed description and interpretation of the fossil record, Bronn derived two concluding "fundamental laws". One was an *intrinsic* 'law' which governed the *progressive* trend that could be clearly discerned in the history of each group of organisms, irrespective of the particular adaptations they displayed; the other was an *extrinsic* 'law' which governed the *adaptive* potentialities of organisms in relations to the environment available. The first law was "positive and productive" in the sense that it accounted for the coming-into-being of genuinely novel and more complex forms of existence; the second was "negative and prohibitive" in the sense that it could only determine which potential forms of existence could actually survive, given the world as it was in the past and as it became in the course of time. These were the fundamental characteristics of the creative force, as observed in its effects; the exact nature of the force remained unknown, but certainly not unknowable.

III

The general form in which Bronn presented his interpretation of the fossil record would not have struck his readers as in any way unfamiliar. It is characteristic of the period that he should have tried to define the workings of the creative force in terms of phenomenological 'laws' rather than directly causal theories. In doing so he was also following a well established tradition in the biological sciences. This tradition was particularly appropriate, and had proved itself of great heuristic value, in the science of embryology. Throughout the nineteenth century the conceptual links between embryology and palaeontology provided a fertile source of analogies. This is reflected in the use of the same terms (evolution, development, *Entwickelung*, etc.) to describe directional changes in organisms both in embryonic growth and in geological time (this has been, incidentally, a perennial cause of historical confusion and misinterpretation)[7]. Both sciences were concerned with understanding the process of coming-into-being of organisms; both, on different levels, faced the problem of explaining the emergence or diversity and novelty in organic form and function during the passage of time—whether the short time-scale of an individual life-history or the immensely long time-scale of the fossil record.

In embryology earlier attempts to explain the extremely complex phenomena of development in crudely causal-mechanistic terms had been abandoned as sterile; and attention had been turned toward more precise description of the actual modes of development, in order to extract phenomenological 'laws' or regularities by which to characterise it. The most distinguished and influential example of this tradition was Karl Ernst von Baer's definition of four 'laws', which in effect summarised embryonic development as a process of gradual differentiation of the more specific from the more general[8]. With superbly detailed description, von Baer had effectly demolished the older belief that the individual organism ascended the Scale of Beings during its development: on the contrary, the major divisions (*embranchenents*) defined by

Cuvier were as distinct in early embryonic stages as in the adult organisms.

It is thus no accident that von Baer, like several other distinguished scientists (for example, Geoffroy, Pander, Agassiz) worked in both embryology and palaeontology; or that Bronn himself expressed his palaeontological results in terms of 'laws' similar to those established in embryology. However much the analogy may have been misused at times, as for example by Chambers, in both fields this concentration on phenomenological 'laws' rather than causal mechanisms was undoubtedly a wise policy for research. It was more important to establish, in the first place, precisely what regularities underlie the changes of form that occur in embryonic development and the changes of fauna and flora that have occurred in the history of the earth, than to attempt causal explanations of these changes on the basis of insufficient factual data.

Nor should this tradition be taken to characterise only those who later expressed reservations about Darwin's theory. On the contrary, the same kind of formulation can be found in a famous paper that Alfred Russel Wallace (1823–1914) sent from Sarawak in 1855. In this brief essay Wallace argued that the facts of geographical and geological distribution of species suggested the 'law' that "Every species has come into existence coincident both in space and time with a pre-existing closely allied species". Very probably he was already convinced that some kind of trans-specific evolution was causally responsible for this 'law'. However there is no reason to infer that his use of 'law' and of the terminology of 'creation' were devices to conceal his true beliefs: on the contrary he was doing no more than many of his scientific contemporaries (including Bronn), in trying to circumscribe the nature of the origin of species by defining the character of 'creation' in terms of phenomenological regularities. Such discussions of the problem were not taboo for religious reasons: at least among scientists it was generally assumed that 'creation' must be a natural process of some kind. All that was suspect in the scientific community was undisciplined speculation about this process. In 1855 Wallace knew he had no causal mechanism to propose and therefore suggested none: but when, three years later, the idea of natural selection suddenly came to him (as it had to Darwin independently nearly twenty years earlier) he had no hesitation about sending a paper on the subject back to England for possible publication in one of the scientific periodicals. It was this second essay on the problem which, turned into a joint paper with Darwin, was read to the Linnean Society in 1858 and thereby became the first public exposition of the theory of natural selection[9]. The fact that it then created no stir is commonly taken to show the intellectual lethargy or obscurantism of the biologists of the time; but alternatively it could be regarded as a sign that such arguments about the mode of origin of species were already familiar, and that it was not immediately obvious from the bare case presented in the paper that this new hypothesis was any more plausible than others had been. In any event, however, it was the arrival of Wallace's paper on natural selection that caused Darwin to shelve his slow preparation of a massive work with that title, and to compose hastily the persuasive 'abstract' that was published in 1859 as the *Origin of Species*.

IV

Before considering the way in which Darwin handled the evidence of palaeontology, it is worth summarising the main features of Bronn's synthesis, for this represents accurately the general state of palaeontological opinion at the time. Like Lyell and Cuvier before him, Bronn and most other biologists of the 1850s believed in the objective reality of the species as a unit in natural history, partly because there was no positive evidence for variability beyond narrow specific limits, but more fundamentally because any other conclusion would have conflicted with their strong sense of the 'adaptedness' of every species to its appropriate mode of life. This was sometimes—but not always—expressed in the traditional language of 'design'; but it was grounded quite as much in the biologists' experience of the astonishing complexity of adaptation as in the desire to use the evidence of design for purposes of natural theology. To believe that such intricately coordinated organic mechanisms had come into being by 'chance' or 'accident', as theories such as Geoffroy's were felt to imply, was literally inconceivable: whatever the creative force might be, Bronn and his colleagues felt it must certainly include the power of regulating the precise adaptation of each newly produced species. Once produced, that species would necessarily be limited in its variability, for outside those limits it would no longer be viable.

Bronn accepted Lyell's position (or more accurately, both of them adhered to the same consensus of opinion) not only on the reality of species but also on the gradualistic and piecemeal manner of faunal and floral changes in the course of geological time. From this it followed necessarily that both interpreted all apparently abrupt changes between formations as purely accidental. Such changes were not due, as a dwindling minority of scientists such as Élie de Beaumont and Agassiz believed, to genuinely sudden events on a wide or even global scale: they were merely the natural result of the local absence of strata representing intervals of time during which the organic world had changed in the usual piecemeal manner.

However, on the *degree* of imperfection of the fossil record Lyell and Bronn were sharply divided. Because Lyell still clung desperately to his original belief in the steady-state history of the organic world[10], he was obliged to maintain with equal tenacity his view that the fossil record was extremely imperfect—so imperfect that any appearance of 'progress' in the history of life was purely an illusion. However this view was becoming less and less tenable every year. Ironically it was the vindication of Lyell's belief in slow piecemeal organic change that was chiefly responsible for undermining the credibility of his belief that there had been no 'progress' in the history of life: every discovery of a new formation, with a fauna and flora that could be intercalated into the succession previously known, was further evidence not only for gradual organic change but also for the relative perfection of the fossil record. Of course Bronn and other palaeontologists were as well aware as Lyell of the intrinsic imperfections of the record: of the whole groups of organisms that were only preserved under exceptional circumstances (for example insects) or never preserved as fossils at all. However this did not affect their increasing confidence in the adequacy of the fossil record as evidence for the major outlines of the history of those groups which possessed readily fossilisable skeletal parts. In other words they recognised that the fossil record was very far from perfect, but that within definable limits it was becoming yearly more reliable as a *sample* of the past history of life.

This sample seemed to show unambiguously that there *had* been certain kinds of 'progress' in the history of life. Within some groups (notably the vertebrates), classes with increasingly 'high' and complex organisation had come into being successively in the course of time; in many more groups there had at least been an increasing taxonomic and adaptive diversity. On the other hand there were definite limits to this 'progress': in particular the earliest known forms of life already clearly belonged to one major group or another (like the early stages of von Baer's embryos) and were in no obvious sense 'imperfect' or poorly adapted relative to later forms: there was no evidence of their gradual development from rudimentary 'monads'. But if Lyell was gratified by this confirmation that any Lamarckian theory was more untenable than ever, he must have been disconcerted to find the same geological research on the earliest strata telling more and more against his hypothesis of the destruction by metamorphism of the earliest 'chapters' of the history of life. The reality of metamorphism as a concomitant of igneous and tectonic activity had become accepted in geology; but the unmetamorphosed sediments that Murchison, Sedgwick and others had been describing were cumulative evidence that metamorphism alone could not explain the dwindling of the fossil record when traced back into the earliest Palaeozoic strata. This seemed to be genuine evidence that the beginnings of life on Earth had been preserved as adequately (though far from perfectly) as later periods of its history. Lyell was reduced to ever more elaborate intellectual acrobatics in order to explain this evidence away, and to re-assert the extreme imperfection and unreliability of the fossil record; but to Bronn and most other palaeontologists no such acrobatics were necessary, and the record could be read straightforwardly as a book (the favorite metaphor for writers on all sides) which was certainly defective in parts but not so fragmentary as to be wholly misleading.

On this more optimistic estimate of the fossil record it followed, as already implied, that Lamarckian theories of gradual transmutation of species were definitely without foundation. There was no fossil evidence of one species changing gradually into another when traced through successive strata; and—much more seriously—there was no fossil evidence that any of the major groups, with their distinct types of anatomical organisation, had had any common ancestors.

This is not to say, of course, that there was no evidence from other sources to suggest 'affinities' of some kind between different classes of organisms. On the contrary, ever since Cuvier systematised the study of comparative anatomy, it had been well known that the different classes within each *embranchement* (for example mammals, birds, reptiles, etc., within the vertebrates) had many features of anatomical construction in common, and that these similarities were independent of the adaptive modifications of individual species. It is easy in retrospect to see how the existence of such homologies could be turned into evidence for a common evolutionary origin of the classes concerned; but at the time this conclusion was far from self-evident, because Owen's alternative explanation in terms of archetypes was generally felt to be altogether more satisfying.

V

This was the somewhat unpropitious intellectual climate in which Darwin had to attempt to convince the scientific community that an evolutionary theory was not merely respectable scientifically but even plausible. By the mid-1850s Darwin himself was known and respected as a distinguished naturalist, a geologist of some originality, and a competent systematic biologist. His voyage with the *Beagle* twenty years earlier had been chiefly important for giving him vivid first-hand experience of the kinds of large-scale geological and biological phenomena that were exercising the minds of most naturalists of the time: the problems of mountain-elevation, for example, and those of biogeography. His own observations had been careful and competent rather than original, and there is little evidence that his thoughts on species were particularly unorthodox. One of his few original contributions to scientific debate arising from the voyage had been his highly Lyellian interpretation of coral reefs and atolls; and it was as a geologist of some promise that he first became known in scientific circles after his return to England.

While subsequently writing up the scientific results of the voyage for publication, however, he became increasingly aware of the centrality of the species problem, and began, in a famous series of notebooks, to speculate freely on the matter (a method that contrasts strangely with his much later reminiscence that his approach had been one of dispassionate 'Baconian' fact-collecting)[11]. Once he had begun to question the conventional belief in the intrinsic limitations of intra-specific variation, it was possible for him to see the relevance to animal and plant species of Malthus's much earlier comments on the interactions of food supply and reproduction rates in human populations[12]. By 1842 his hypothesis of 'natural selection' in the wild, directly analogous to the artificial selection exercised by breeders to improve their plant and animal stocks, was ready to be sketched out as the central postulate of a theory of the causal *mechanism* of species production[13].

Even in this first sketch, his earlier geological work played a crucial part. His hypothesis of the production of new species from varieties and races was analogous to his interpretation of fringing reefs and atolls as different stages in a single gradualistic process; but even more significantly he used his uniformitarian hypothesis of continental elevation and subsidence as a device for sidestepping Lyell's own objection to Lamarckian transmutation. Lyell had argued that even if (which he doubted) there were no intrinsic limits to variation, a changing environment would still not result in species transmuting into new forms, because they would be extirpated by the invasion of other species already better adapted to the new conditions. Darwin, however, perceived that this would not apply on a continent emerging gradually from below sea-level. While the highest points on such a continent were emerging in the form of gradually enlarging island areas, their initially restricted faunas and floras would be presented with an ever-increasing diversity of habitats; yet their geographical isolation would protect them from competition with species already adapted to these new habitats. Hence there was "no point so favourable for generation of new species" as such an island area. Conversely, a continent gradually sinking below sea-level (leaving fringing reefs and atolls as signs of its subsidence, if situated in tropical latitudes) would be the *least* favourable areas for speciation.

The Lyellian steady-state form of this hypothesis then gave Darwin an explanatory bonus: for he believed that the chances of fossilisation of terrestrial organisms were lowest in the areas of elevation, where the speciation would be occurring, whereas the conditions in which such organisms were most likely to be preserved were those of the areas of subsidence, where speciation would *not* be taking place. This argument seemed to dispose neatly of one of the most formidable objections to any hypothesis of gradual transmutation: for under the circumstances postulated it was not surprising that direct fossil evidence of transmutation was conspicuously lacking; indeed, Darwin wrote, it would be "wonderful if we should get transitional forms" preserved at all.

Darwin was clearly worried even at this stage by the failure of palaeontology to provide support for an evolutionary theory: "if views of some geologists be correct," he warned himself, "my theory must be given up". The views referred to were simply the general opinion that the fossil record, for all its obvious local gaps and intrinsic imperfections, was a sample of the history of life that was becoming yearly more reliable and (relatively) complete. In the face of this opinion, Darwin's belief in extremely slow trans-specific evolution forced him into what he termed "Lyell's doctrine carried to extreme", namely into an assertion that the fossil record was more fragmentary and unreliable than even Lyell believed. The fossil record was commonly conceived, as we have seen, as a book in which many pages were certainly missing, but in which further research was gradually filling the gaps and generally improving the text. Lyell had argued, against this confident optimism, that only a few pages in each chapter were likely ever to have been preserved. Darwin was obliged to emphasise even further the fragmentary nature of the record; but characteristically he made a virtue out of necessity, concluding that "if geology presents us with mere pages in [one] chapter, towards [the] end of a history . . . the facts accord perfectly with my theory".

Thus even in the 1842 sketch of his theory, Darwin was already having to rely heavily on negative evidence when considering the palaeontological contribution to the species problem. Two years later, in 1844, he amplified his sketch into a carefully composed essay, which would have been published if (as he feared) he had died before completing a fuller treatise on the subject[14]. Several factors combined to make him slow to publish this fuller version. He was cautious by temperament. He was aware that his theory involved the rejection of several cherished scientific assumptions, particularly that of intrinsically limited variability. He was equally aware that it claimed to account for all the accurately coordinated and designful aspects of adaptation in terms of a mechanism that would seem to depend on chance and accident, and which would therefore be felt to be metaphysically and theologically repugnant, especially when applied to Man. He knew that the theory would therefore carry conviction only if it could be supported by a mass of detailed evidence. Finally, however, all these factors were reinforced by the justifiably scornful way in which Chambers' specutions were received by virtually all the scientists Darwin most respected. His own theory, if it was to persuade the scientific community, would have to be seen to be in an entirely different class: it would need to overwhelm its potential critics by the sheer weight and scientific authority of its arguments.

This authority he could only earn by showing himself to be a thoroughly competent systematist with first-hand knowledge of the practical problems of dealing with intra-specific variation and interspecific distinctions. He therefore embarked on eight years of solid systematic research. However, the group he chose to work on indicates that his intentions went further than merely proving his scientific credentials. The barnacles (Cirripedia) were a highly appropriate group, not only because they had been placed in a 'transitional' position in some earlier classifications of the invertebrates, but also because they illustrated with particular clarity some of the features of organic diversity that would need to be explained by any satisfactory theory of evolution. It was already known that in the early stages of their life-history barnacles were fairly normal crustacean larvae, and that at a later stage they became permanently sessile, converted their legs into an intricate food-collecting device, and lost many of the functions and organs associated with a normal free-swimming life. If this peculiar 'regression' to what might be regarded as a 'lower' or less complex form of existence was to be explained in evolutionary terms, 'evolution' would have to allow for regressive as well as progressive trends. However since the regression was as clearly adaptive as progressive trends in other organisms—barnacles were a manifestly successful group and were superbly adapted to their sessile mode of life—this suggested that a single causal explanation might underlie evolutionary changes towards simplicity as much as those towards complexity. In other words, the barnacles suggested confirmatory evidence of Darwin's hypothesis that the same kind of Malthusian selective pressure must be causally responsible for *both* kinds of change.

This implied a crucial shift in the terms of reference of the whole problem. Lamarck had felt it necessary to postulate an intrinsic 'tendency to progressive improvement' in organisms, modified secondarily by the requirements of the environment. Bronn's two 'fundamental laws' are surprisingly similar on a conceptual level, even though he rejected the possibility of Lamarckian transmutation: as we have seen, he postulated one essentially directional 'law' and another essentially adaptive. Darwin, in effect, was to re-state the problem by suggesting that a single 'law', essentially adaptive in character, might account not only for the adaptive features of particular organisms, but also simultaneously for the existence of both 'progressive' trends in the history of life *and* the occasional 'regressive' instances such as the barnacles.

It is characteristic of Darwin's interest in the palaeontological dimension of the problem that he tackled simultaneously the systematics of both living and fossil barnacles. In his work on the latter[15], a conventionally phrased reference to "the exhaustless fertility of Nature in the production of diversified yet constant forms" gave no hint of the radical character of his unpublished theory. However this reticence was not only due to his policy of scientific caution but also to the plain fact that his fossil barnacles gave no positive evidence for slow trans-specific evolution: on the contrary they confirmed the general opinion that intra-specific variation had definite limits, and that species were indeed "diversified yet constant forms" when traced through geological time (Fig. 5.1).

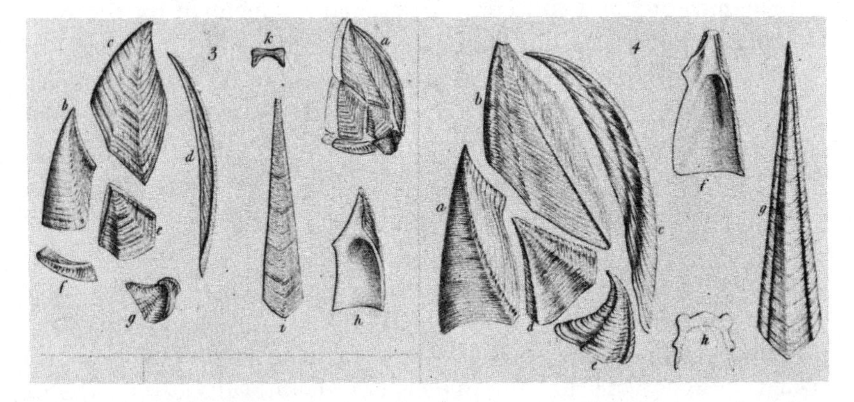

Fig. 5.1. Darwin's illustrations of two related species of fossil barnacles, from his work on systematic palaeontology (1851)[15]*. Note the slightly but significantly different shapes of the homologous plates in the Eocene* Scalpellum quadratum (left) *and the Cretaceous* S. fossula (right). *Darwin referred to such separate species as 'diversified yet constant forms'; he had no fossil evidence for gradual trans-specific evolution, although he was working at this time on his hypothesis of natural selection.*

While Darwin was addressing himself to the empirical difficulties of working with natural species, he was also carrying out work of far greater importance for his theory. He was exploring the implications of his central analogy between artificial and natural selection, by studying the range of variation obtainable by breeding, particularly in pigeons. His own breeding experiments, and his wide correspondence with those with practical experience of breeding, was to bring a whole new area of evidence to bear upon the species problem. But of course its relevance depended on the validity of the analogy. In Darwin's own mind it seems to have been not only valid as a strict analogy, but also methodologically important as the nearest approach to Lyellian actualism that was available. If the transmutation of species was too slow to be observed in operation on a human time-scale, as Lamarck had asserted, then actualism was strictly inapplicable. However Darwin felt that artificial breeding experiments could serve as an accelerated replica of what must have happened far more slowly under natural conditions. This belief, however, depended on taking the analogy very seriously, and assuming that 'Nature' (which Darwin's language tended to personify) did 'select' in ways directly comparable to the conscious agency of a human breeder[16].

VI

Darwin began to write his massive treatise on *Natural Selection* in 1856. But as we have seen, in this task he was overtaken by events, and he hurriedly brought out an 'abstract' *On the Origin of Species by Means of Natural Selection*[17]. The fuller work remained unpublished, though some portions of it were incorporated into his later books on particular topics.

The *Origin* was in effect an amplified version of Darwin's 1844 essay, and it retained from that early draft a distinctive structure of argument and balance of contents. Out of thirteen chapters (excluding a final summary of the whole book) no less than eight were devoted to the variability of organisms under domestication and in the wild, to the analogy between artificial selection and selection by 'Nature', and to related questions at the species level. Furthermore these were the *first* eight chapters, the heart of the book, and they were clearly intended to establish the plausibility of natural selection as a mechanism for trans-specific evolution. Only in the last five chapters did Darwin turn to consider briefly the implications of applying this micro-evolutionary theory to the large-scale evidence of the fossil record, of biogeography and of comparative anatomy and embryology.

Of these 'accessory' topics the first was still by far the most worrying from Darwin's point of view. He was well aware that palaeontology, by the intrinsic nature of its material, could alone supply direct and positive evidence that varieties and races really *had* given rise to new species in the course of time, or that evolution at a higher level than the specific could really account for all the diversification observed in the history of life. But as we have seen, it was precisely this positive evidence that palaeontology did *not* supply. As foreshadowed in the earlier drafts of his theory, Darwin was therefore obliged to fall back on an ultra-Lyellian argument for the extreme imperfection of the fossil record. He was forced to rely heavily on negative evidence to explain the absence of intermediate forms between similar species, and the still more embarrassing absence of intermediate forms between the major classes of organisms. Such arguments had had some reasonable plausibility in 1842; but by 1859 the progress of palaeontological research, particularly on the early Palaeozoic strata, had made Darwin's position look suspiciously like a case of special pleading.

His argument was more persuasive in the interpretation of biogeography; and on comparative anatomy and embryology it was bound to seem convincing to any reader who found the 'metaphysical' character of Owen's archetype interpretation unsatisfying, and who was therefore pre-disposed to look favourably on an evolutionary explanation of anatomical diversity. Fundamentally, however, the prospects of the theory depended on the credibility of the first part of the book—and particularly on the validity of the analogy between artificial and natural selection—and on the acceptability of Darwin's sweeping use of an imperfect geological record to by-pass the lack of positive fossil evidence for extremely gradual trans-specific evolution.

The unusual circumstances in which the *Origin* was published gave Darwin's theory one great advantage which it might not otherwise have had: it was short, relatively untechnical, and therefore accessible to a wide reading public and not just to the more professional scientific community. Had the theory been published in its originally intended form it is doubtful whether it would have had such an impact on currents of thought outside biology. On the other hand, the form in which it was in fact published was understandably disappointing to Darwin's fellow-scientists. The style of his work was even more that of an advocate than Lyell's had been; but the persuasive imaginative lines of thought that the reader was invited to follow were not supported, as Lyell's were, by comprehensive and detailed evidence or by full citations of the work of other scientists. Yet on the other hand the *Origin* claimed to be a work of original scientific research and not primarily a work of popularisation.

Darwin's fellow-scientists were not so churlish, however, as to reject his argument on the grounds that its form of publication was unconventional. Even if they regretted the lack of detailed documentation, and looked forward to the publication of the fuller work, they still recognised the importance of Darwin's hypothesis. Bronn, for example, saw at once that it deserved to be considered as widely as possible by the scientific community; and he therefore made it available in the chief scientific language of the time, publishing the first German edition as early as 1860. Bronn realised that Darwin was proposing, in effect, that 'natural selection' was the mechanism through which the 'creative force' could have operated. While he was not convinced that this mechanism was wholly satisfactory, it was certainly the *kind* of hypothesis he could approve whole-heartedly; he saw clearly that it was an important contribution to the continuing debate about the character of the natural agency of 'creative force'.

The theory was viewed in the same light by another distinguished palaeontologist, François-Jules Pictet (1809–1872)—Darwin's contemporary—whose review of the *Origin* in 1860, in a widely read Swiss periodical, brought the work to the attention of the French-speaking world (the book itself was published in French in 1862). "It is a long time since we have read anything more comprehensive or more interesting on this difficult and controversial question", he wrote, praising Darwin's work particularly because the problem was presented "in a new form, and, to some extent, detached from the ordinary routine" of evolutionary speculations. In other words the first part of the *Origin*, by concentrating on the variability of living species and introducing the concept of natural selection, had brought new areas of evidence to bear on the problem[18].

Pictet found Darwin's arguments highly persuasive—but only up to a certain point: "a moment came when his imagination moved more swiftly than mine, and where he has drawn some conclusions which seem to me not in accord with the facts acquired". Pictet was prepared to believe that transient varieties could indeed become permanent new species under the influence of natural selection—he himself had expressed the probability of some mechanism of inter-specific evolution long before, in his highly respected *Treatise on Palaeontology* (published in the same year Darwin wrote his draft essay)[19]. But then there was "a sudden jump" in Darwin's book, "by which the reader is asked to pass from the careful study of facts to the most extreme theoretical consequences". "An inflexible logic" drove Darwin to extrapolate his hypothesis from the level of inter-specific change—which Pictet readily accepted—all the way to the evolution of every form of life from a few or even a single original type. Such an extreme extension of the theory raised grave difficulties. How could organs of fundamentally new function be formed by gradual transitions—lungs from gills, or wings from forelimbs—without fatal loss of viability in the intermediate stages? Indeed, the more one stressed (as Darwin did) the struggle for existence, the more acute this long-standing objection was bound to be. Yet Pictet agreed that Darwin had marshalled strong indirect evidence for the common origin of all the species within each of the larger taxonomic groups: the theory had great explanatory value when applied to some of the evidence of comparative anatomy, embryology and even palaeontology. This posed a curious dilemma: a theory so attractive on some grounds was scientifically unacceptable on others; yet what better alternative could be proposed? "Here", said Pictet, "I feel myself too weak, and very near to *I don't know*, the usual conclusion to these mysterious questions". Nevertheless, he restated tentatively his own view—that the "force" of "normal generation", whose effects Darwin's work on variability had substantially extended, must necessarily be supplemented from time to time by the action of a "creative force", which alone could account for the apparently sudden production of fundamentally new types of organisation. Although the nature of this second force remained unknown, Pictet emphasised (like Bronn) that it must be as *natural* as other forces in the physical world. He argued that even Darwin himself was reluctant to postulate the common origin of all the major groups of animals, and *a fortiori* the common origin of animals and plants; so that even Darwin was implicitly accepting a limited role for Pictet's creative force, and the disagreement between them was really one of the relative roles to be assigned to the two agencies.

Pictet's review illustrates the typical reaction of palaeontologists—and many other biologists—to Darwin's theory. It was plausible as an explanation of small-scale change; but the larger the scale on which it was applied the more difficult it was to believe. It was relatively simple to imagine that intra-specific variation, acted on by natural selection, might give rise to new species of fairly similar form and habit; it was far harder to conceive how the fundamentally different anatomical and physiological organisation of the major groups of organisms could have originated by the same means, however many millions of years were conceded for the purpose. Still more seriously, the fossil record failed to provide any evidence of any such gradual transitions.

This last point was taken up by John Phillips, who was by now one of the leading palaeontologists in Britain and had risen to be Professor of Geology at Oxford. In 1860, while he was also President of the Geological Society, he gave the Rede lecture at Cambridge on *Life on the Earth: its Origin and Succession*, in which he reviewed the current evidence of palaeontology in the light of Darwin's theory[20]. Most palaeontologists of the time would probably have agreed with him when he maintained that Darwin had grossly over-stated the case for the imperfection of the fossil record. Imperfect it certainly was, but in broad outline, and particularly for shell-bearing marine animals, it was good enough to test the plausibility of Darwin's belief in extremely slow trans-specific changes. Not only was there no positive fossil evidence for such transitions, but much more seriously it was now clear that the earliest known forms of Palaeozoic life were already highly complex organisms; and the absence of simpler precursors could *not* any longer be attributed to the metamorphic character of the earliest rocks. Phillips' review of the fossil record shows clearly that he was well aware of the vast scale of geological time; but this, he felt, did not justify Darwin in multiplying that time-scale extravagantly just in order to accentuate the gaps in the record and evade the difficulties presented by it (Fig. 5.2).

VII

The melodramatic character of a famous episode at the British Association meeting in Oxford in 1860, when Darwin's friend Thomas Huxley (1825–1895) attacked Samuel Wilberforce, has tended in retrospect to become a symbol of the Darwinian debate seen as a simple conflict between science and religion. In fact it is difficult to discover precisely what happened on that occasion, since most of the accounts date from twenty or thirty years after the event, and are evidently coloured by the much cruder ideological divisions that had by then developed. What seems clear, however, when the episode is shorn of its mythological element, is that the then Bishop of Oxford's facetious remarks about the theory of evolution were as embarrassing to most of the scientists present as they were offensive to Huxley, who leapt to Darwin's defence. In the development of the Darwinian debate it is important to distinguish the reactions of the scientific and scholarly community from reactions at a less sophisticated level[21]

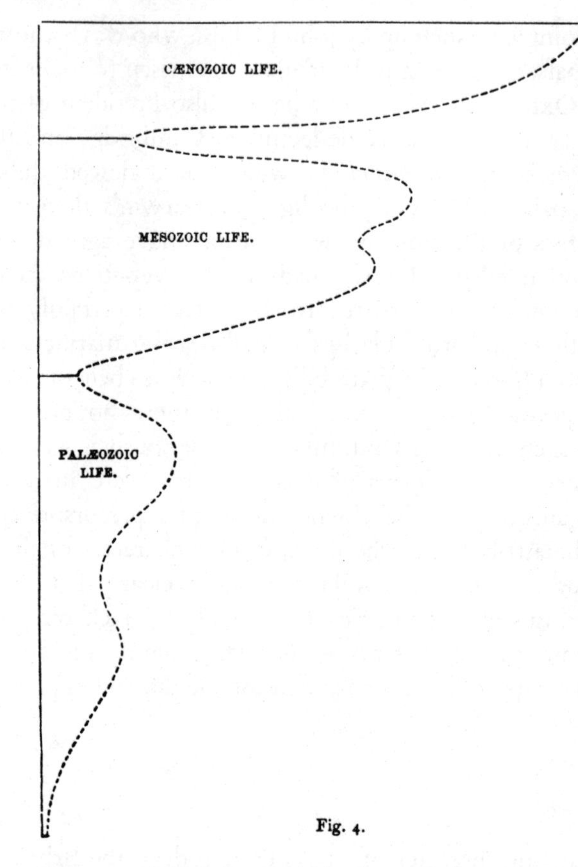

Fig. 5.2. Phillips' diagram of the fluctuating but progressively increasing diversity of life during Earth-history (1860)[20]: major phases of change delimited three great eras in the history of life. The relative lengths of the eras were estimated from maximum thicknesses of known strata, and were thought to total nearly 100 million years (this is five or six times shorter than the modern radiometric estimates, whereas Darwin's estimates were five or six times longer).

Much of the supposed religious opposition, for example, was the conservative reaction of ordinary clergy and religious people who had been sheltered by the intellectual isolation of Britain from the rapid development of critical theology on the Continent. Darwin's *Origin* was generally regarded as merely a part of a much wider threat to the literalism of a comfortably ensconced orthodoxy, a threat that was embodied much more substantially in *Essays and Reviews* (1860).

Without this volume, which at last brought scholarly German theology forcibly to the attention of the English reading public, Darwin's work might have had few immediate repercussions outside scientific circles.

This is not to say that there were no metaphysical or theological components in the opposition of scientists such as Richard Owen, who wrote one of the first and most important critical reviews of the *Origin* (and who is said to have coached Wilberforce for his British Association speech)[22]. But Owen's criticism of Darwin's theory was not a defence of biblical literalism or of special creation. What was at stake was the *nature* of the evolutionary process, and in particular its designful character. It is almost impossible to discover how far Owen's strong sense of the 'plannedness' of the organic world was grounded in theological conviction and how far in biological experience: the distinction is perhaps ultimately meaningless. But whatever its roots in the mind of Owen—and of most of his scientific contemporaties—it is clear that it led them to oppose Darwin's *mechanism* for evolution rather than the idea of evolution itself. What was unsatisfying and even repugnant about Darwin's theory was that it claimed to account for all the 'plannedness' of the organic world—both the intricate adaptations of particular organisms and the grand drama of the whole history of life—in terms of the interaction of chance variations and haphazard environmental changes.

When in 1860 Owen published a general work on *Palaeontology*, therefore, he explicitly accepted the probability of "a continuously operative secondary creational power" as the agency responsible for the evolution of diverse organisms in the course of time[23]. Owen's language is reminiscent of Bronn's and Pictet's, with whose work he was almost certainly familiar; none of them were in any doubt that the causal agency involved was 'secondary', that is, *natural*. But Owen emphasised the need to distinguish clearly the proposition of "species being the result of a continuously operating secondary cause" from the question of "the mode of operation of such creative cause". The first could be accepted (as Owen himself did) without necessarily being convinced by the adequacy of any current versions of the second. The version proposed by Darwin and Wallace was at present inadequate, in Owen's view, primarily because "observation of the actual change of any one species into another . . . has not yet been recorded". In other words, Owen was beating Darwin with a methodological stick provided by Lyell: and he was beating him at the most vulnerable point, namely over the lack of fossil evidence for evolution. Darwin had of course tried to cover that point by appealing to Lyell's emphasis on the vast scale of geological time and the imperfection of the fossil record, but this was bound to seem like special pleading.

Lyell's own position relative to Darwin's theory was—not unnaturally—highly complex. Several years earlier he had begun to reconsider his original views on the stability of species, and it was largely the weight of his opinion that had persuaded Darwin to publish his theory[24]. Darwin had been deeply influenced by Lyell's methodology; and yet Lyell was naturally reluctant to espouse any evolutionary theory—even one by Darwin—for this would seem to imply that the progressionists had been (at least partly) right after all, and Lyell himself wrong in holding so doggedly to a steady-state view of the history of life. It was not until 1868 that Lyell publicly came out clearly in favour of Darwin. Even then the relative lukewarmness of Lyell's support was a disappointment to Darwin—though in fact Lyell deserves great respect for making such a difficult *volte face* publicly when already in his later sixties.

It is significant that Lyell's first major statement on Darwin's theory was included, almost as a digression, within his semi-popular book on the *Antiquity of Man*[25]. The palaeontology of the human species had only recently re-emerged into the forefront of scientific debate, chiefly as a result of the excavations at Brixham cave in Devonshire in 1858. Unlike most of the caves excavated earlier, the evidence for the co-existence of tool-making primitive man and the extinct Pleistocene mammals was here incontestable. Almost at the same time, the validity of a 'Palaeolithic' period in human history was confirmed by the belated vindication of claims that had been made for many years by a French Customs official and amateur archaeologist, Boucher de Perthes, to have found flint implements *in situ* in the bone-bearing Somme gravels (Boucher's claims had been treated with extreme scepticism by the scientific community, probably because he used them explicitly to support a—by then—antiquated diluvial theory). Using the now well-established methods of stratigraphical palaeontology, the Palaeolithic had even been subdivided, as early as 1860, into several sub-periods in which primitive Man had been associated with different mammalian faunas. Archaeological work had meanwhile been throwing new light on the more recent 'Neolithic', 'Bronze Age' and 'Iron Age' cultures[26].

This mass of research was highly important, because it extended the history of man unambiguously back into the lengthy geological time-scale, and showed that there had been ample time for the very gradual differentiation of the various races of mankind and for the slow development of civilised societies. (The political implications of this conclusion were far from negligible, for it helped to undermine contemporary theories of the separate origins of the various races—theories which, then as now, were being used to give 'scientific' respectability to racialist attitudes.)

However, while Lyell could sketch out a 'evolutionary' history for the human species itself, he could not give any positive evidence of its origin. Human skeletal remains had actually been found in association with extinct mammals as early as the 1830s, but they were well within the range of variation of the living human species. The first fossil primates had been found about the same time; but with one exception no fossil hominids were known. The exception was the first Neanderthal skull, discovered in 1856; but the geological age of this fossil was highly uncertain, and when Huxley studied it he was obliged to conclude that it was *not* an intermediate between man and the anthropoids —a conclusion he may have reached with some inward reluctance, since he had become one of Darwin's first and most enthusiastic supporters[27]. (The first fairly plausible intermediate, *Pithecanthropus* or 'Java Man', was not found until the 1890s).

Despite the absence of positive evidence for the origin of man, however, Lyell felt able to insert a section on the general question of evolutionary theory, with the clear implication that it might ultimately be found applicable to man himself. This application of the theory, which Darwin himself wisely had only hinted at in passing, was of course the emotive factor behind much of the Darwinian debate, including Huxley's public confrontation with Wilberforce. But it is important to note that the issue really at stake was not the origin of the human species as such, but rather the status of mind and consciousness, and the place of the moral realm, in relation to the material world. It was therefore possible to argue for the evolutionary origin of the human species, while doubting whether the evolution of mind could be fitted into the same explanatory framework. Obviously this involved applying a Cartesian dualism of mind and matter to the situation, but it was an attitude that had become sanctioned by heuristic success in biology ever since the time of Descartes himself. In other words, by including the study of man within Natural History only in 'bodily' ways (for example in comparative anatomy and physiology) it had been possible to study the nature and history of man as part of the organic world without getting bogged down in the metaphysics of the human mind. Of course the penalty for this compartmentalising of knowledge was very great, and the human sciences are still suffering from its effects; but that should not blind us to its historic advantages.

In explaining his own conversion to an evolutionary viewpoint, Lyell followed Darwin in making a virtue out of necessity in accounting for the lack of positive fossil evidence for the slow transmutation of species. The extreme imperfection of the fossil record explained all. However, Lyell then landed himself in the curious position of believing so strongly that palaeontology *could not* provide evidence for evolution, that he failed even to consider what fragmentary evidence it might conceivably yield. Thus in explaining away the absence of intermediate links, he only cited the difficulties of systematists in deciding the boundaries of fossil species (using prominently Thomas Davidson's superb work on brachiopods). Yet this was no more than the argument Lamarck had used long before. Lyell failed to suggest that a search might be made for *successions* of fossil species occurring in an order that would be meaningful in evolutionary terms, even if the succession was too incomplete to show trans-specific changes in detail. Similarly he mentioned recent discoveries of fossils preserved in exceptional circumstances, for example Oswald Heer's work on the Miocene insects and plants of Oeningen in Switzerland, while ignoring the fact that such chanciness of preservation did not apply equally to all groups. He also cited recent discoveries of fossil faunas that tended to fill up previous gaps in the record, for example the rich Upper Triassic fauna of St Cassian in the Dolomites, filling the gap between the Muschelkalk and the Lias; but instead of interpreting this as a sign that the fossil record was becoming progressively better known and less imperfect, he only emphasised how much remained to be discovered and how little direct evidence for evolution could therefore be anticipated from palaeontology. Most ironically of all, he mentioned the recent discovery of the reptile-like bird *Archaeopteryx*, but only to prove that birds existed much earlier than had been suspected, and hence to underline once more the virtual uselessness of the fossil record: he could not see, as Huxley did soon afterwards, that this famous fossil provided valuable evidence for gradual evolution even across the boundaries of separate classes.

VIII

As so frequently happens in the course of scientific controversy, this potential stalemate between Darwin's hypothesis and the palaeontologists' demand for positive evidence was avoided by a shift in the terms of reference within which the evidence was sought. Darwin had proposed a mechanism for very slow trans-specific change; the palaeontologists replied that no fossil evidence supported such a process. Darwin trusted in the long-continued action of such small-scale changes to achieve cumulatively the evolution of even the highest levels of organic diversity: the palaeontologists replied that the major groups (the *phyla* of later biology) were already distinct in the oldest fossiliferous strata. However, between the level of species and that of phyla there was an intermediate level of taxonomic and morphological diversity, which offered much more promising ground for testing the validity of a general evolutionary theory. The fossil record might be too incomplete in detail to record trans-specific changes, and perhaps still too poorly known in the oldest rocks to record the origin of the phyla. But it was generally agreed to be complete enough to record in broad outline the course of the history of life; and it therefore offered some hope of showing whether different organisms, traced at the generic or familial level, had succeeded one another in the kind of sequence that would be expected on an evolutionary theory.

One of the most enthusiastic exponents of this approach was the distinguished French palaeontologist, Albert Gaudry (1827–1908). Between 1855 and 1860 the Academy of Sciences in Paris had sponsored large-scale expeditions, in which Gaudry took part, to exploit an earlier discovery of rich bone-bearing deposits of Miocene age at Pikermi in Greece. This was a discovery of crucial importance, because it provided one of the first rich faunas of fossil mammals intermediate in age between the two Cainozoic faunas hitherto best known—the Eocene and Pleistocene faunas that Cuvier had first begun to reconstruct. When Gaudry came to write up the results of the expeditions, he said that "the constant purpose" of his research had been to study the nature of "intermediate forms" in fossil mammals[28]. With a degree of historical sympathy that is rare in working scientists, he stressed that the Cuvierian emphasis on the distinctness of species had been essential to the progress of palaeontology; but now, he argued, the underlying unity indicated by the existence of intermediate forms was revealing a still more sublime view of the history of life. Indeed, the discovery of "close links" (*liens intimes*) between the species or genera of successive epochs constituted the most important *testable criterion* for a general theory of "filiation" (that is, evolution), which Gaudry felt in any case was by far the most satisfactory explanation of the fossil evidence.

The deposits at Pikermi provided Gaudry with abundant examples of intermediate forms, which enabled him to present the evidence for the 'filiation' of various mammalian families in the form of phylogenetic 'trees' (Fig. 5.3)—some of the earliest examples of this type of visual presentation of evolutionary results (Bronn and Darwin had both used 'tree' diagrams, but only hypothetically). Gaudry knew these 'genealogies' were no more than provisional, and would need amendment in the light of further discoveries; but the important point was that already the clear *tendency* of new discoveries was towards filling the gaps between previously known species and genera. Thus for example the horse genus (*Equus*) of Pleistocene deposits—and of the living fauna—had been curiously isolated anatomically from all other odd-toed ungulates (*Perissodactyla*). However this isolation was ended, or at least greatly reduced, by the discovery at Pikermi of abundant remains of a genus (*Hipparion*) of less specialised horse-like mammals (Fig. 5.4). Such forms were not only intermediate in anatomy, but were demonstrably of the correct geological *age* to fit into an evolutionary series.

Gaudry was aware, moreover, of the importance of intra-specific variation in fossils. Owing to the richness of the Pikermi deposits he was able to demonstrate the wide but continuous field of variation within what was clearly a single biological species of *Hipparion:* and he commented that such variants might easily have been given separate specific names if they had not been found all together. In other words, the more fossil material available, the more the apparent gaps between the species of different epochs would tend to be closed up.

With such views as these, Gaudry might seem to be logically close to Darwin's position. Yet like other palaeontologists of his generation he distinguished clearly between the 'whether' of evolution and the 'how'; and while believing in the reality of evolution and of the phylogenies he constructed, he felt there were grave objections to Darwin's hypothesis about its mechanism. His reasons for doubting natural selection are typical: it implied the centrality of a 'struggle for existence' and, ultimately the rule of chance and accident; whereas Gaudry's

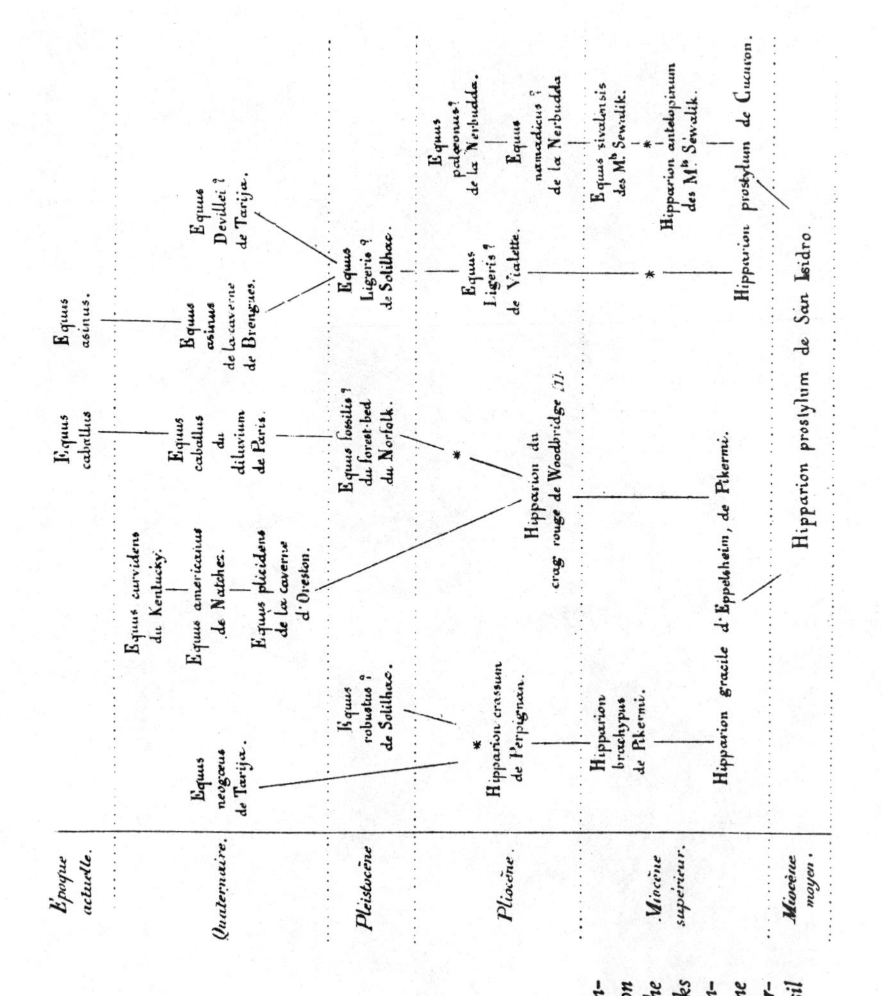

Fig. 5.3. Albert Gaudry's tentative reconstruction (1866) of the branching evolution of modern horses Equus *(top) from the mid-Tertiary horses* Hipparion*; asterisks mark the more doubtful evolutionary connections*[28]*. Gaudry's diagrams were some of the first in which evolutionary interpretations were applied in detail to fossil material.*

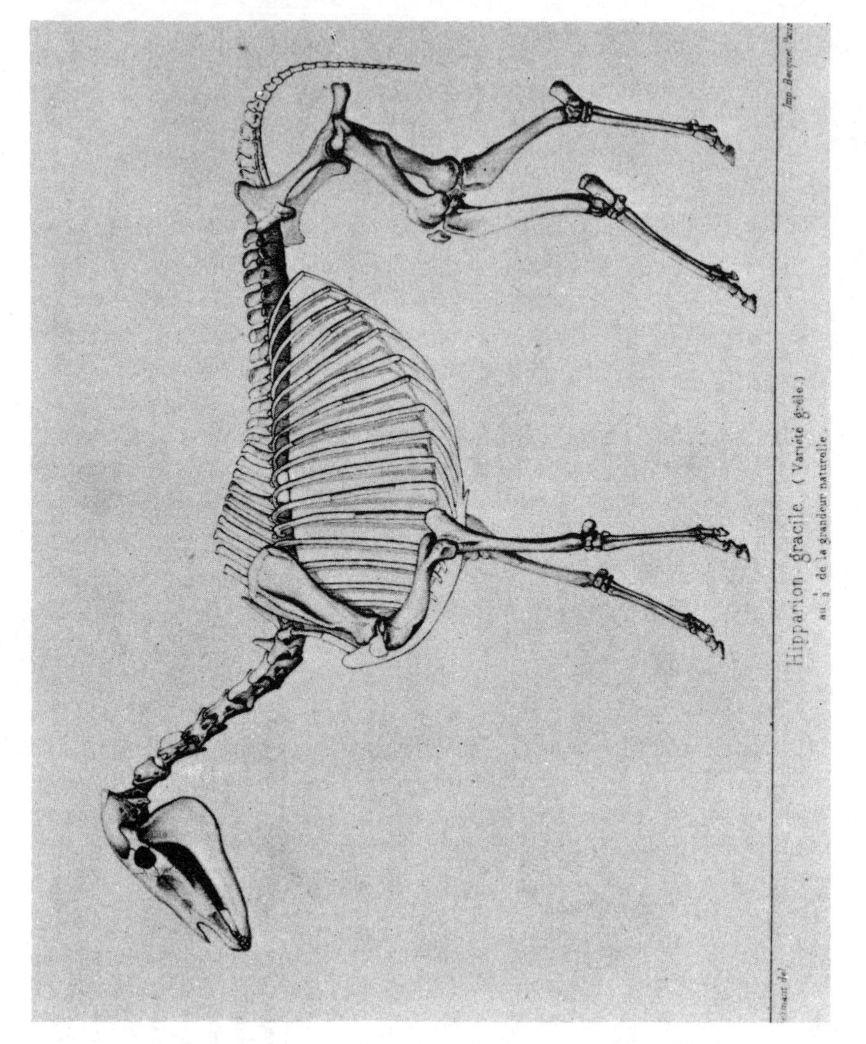

Fig. 5.4. Albert Gaudry's reconstruction (1862) of the skeleton of the Miocene fossil horse Hipparion, *which he analysed functionally and ecologically and interpreted as an evolutionary ancestor of the modern horse*[28].

own ecological reconstruction of the Pikermi fauna and flora showed on the contrary that the world of that epoch, like the present, had been one of ecological balance and harmony. Gaudry's feeling about Darwinian evolution indicates how pervasive was the objection to natural selection on account of its supposed 'chanciness'. With the reinforcement of natural theology, the science of ecology had grown up with an emphasis on the balance and harmony of nature, whereas Darwin's work appeared to replace this picture with one of ceaseless struggle and disharmony (it has needed the work of post-Darwinian ecologists to reconcile these opposites in a new synthesis that can once more emphasise the balance of nature).

Darwin's friend Huxley was, indeed, one of the very few scientists with palaeontological experience to accept not only the general theory of evolution but also Darwin's hypothesis of its mechanism. Huxley was able to bring to the debate much the same range of expert knowledge as Owen; but finding no satisfaction in Owen's theory of transcendental archetypes as an explanation of organic diversity, and clearly preferring a hypothesis that involved an unambiguously physical causal agency, Huxley was a ready convert to Darwin's viewpoint. With his background in comparative anatomy and his experience in palaeontology, he was well placed to follow up the implications of Darwin's theory in these directions and to deploy the latest evidence in its service.

Huxley saw clearly that Lyell's extreme scepticism about the value of the fossil record was unjustifiably pessimistic, and that the tendency of new discoveries to fill previous faunal and floral gaps gave encouragement that they might simultaneously provide a growing body of evidence in favour of Darwinian evolution. He therefore not only welcomed Gaudry's efforts to sketch out the possible cause of evolution at the familial or ordinal level, but also anticipated the possibility of evidence of intermediate forms at an even higher taxonomic level.

The discovery of *Archaeopteryx* provided him with just such evidence at the strategic moment. Ever since the development of lithography early in the century, the finest stone had been worked extensively at Solnhofen in Bavaria; and the fossils from this Jurassic limestone, though not abundant, had become famous for their exceptional quality of preservation. However the first specimen of a feathered bird was found there only in 1861 (a second specimen was not found until 1877). With a collection of other Solnhofen fossils it was purchased in 1862 by the

British Museum, where Owen was now in charge of the Natural History collections. Owen's masterly description of this remarkable fossil concluded that it was "unequivocally a bird", although he admitted that it displayed certain characters known only in the embryonic stages of living birds, and that in general it showed a "closer adhesion to the general vertebrate type"[29]. Shortly afterwards, further study showed that it had toothed jaws; and Huxley realised that Owen's typological language could be readily translated into evolutionary terms. The toothed jaws, long tail and 'embryonic' characters then became simply reptilian traits in what was (Huxley agreed) more or less a true bird (Fig. 5.5).

The Solnhofen quarries had however also yielded a small bipedal dinosaur, *Compsognathus*, which had been described in 1861 by Andreas Wagner, the Director of the Natural History Museum in Munich. Huxley quickly saw that a small bipedal reptile like *Compsognathus*, which also had some remarkably bird-like anatomical features (for example in the pelvis), could go some way towards closing the reptile-bird gap from the other side, even if, as a contemporary of *Archaeopteryx*, it might not itself be in direct evolutionary line. In a lecture to a general audience at the Royal Institution in 1868, Huxley therefore used *Archaeopteryx* and *Compsognathus* as a test case to argue for the plausibility of the evolutionary theory[30]. Significantly, he felt he had to defend the theory against the charge—which was justifiable in view of Darwin's and Lyell's arguments—that it was based on lost evidence and therefore untestable. Using a metaphor appropriate to his property-conscious Victorian audience, Huxley argued that on the contrary the "title deeds" of the theory were as complete and legitimate as the intrinsic nature of the documents allowed. The discovery of *Compsognathus* and *Archaeopteryx*, and their interpretation as a bird-like reptile and a reptile-like bird respectively, showed that there was no insuperable problem in conceiving how even separate classes of animals, with their very different anatomical and physiological organisations, could have had common ancestors. Once that possibility was conceded, then all the formidable body of evidence from comparative anatomy and embryology, and from the geographical distribution of organisms, also fell into place within an evolutionary framework of explanation.

Huxley's line of argument illustrates how the progress of evolutionary re-interpretation within the scientific community was essentially a gradual affair, a matter of cumulative circumstantial evidence, of a

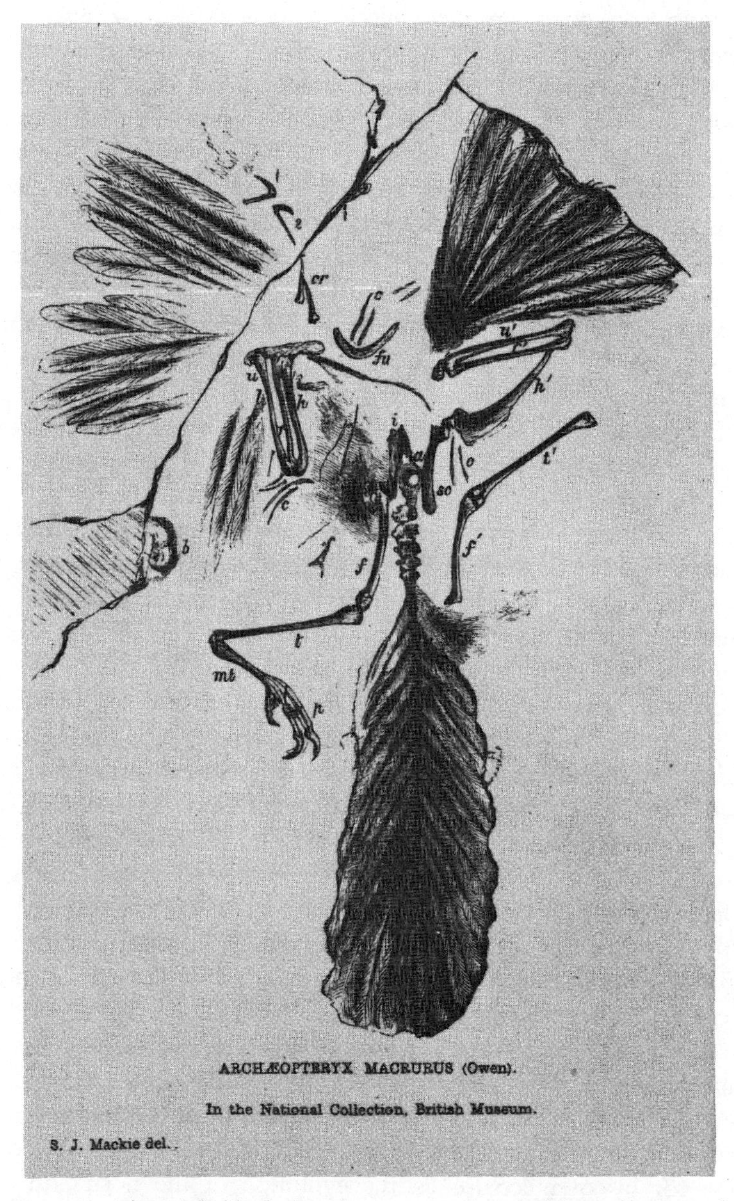

Fig. 5.5. The reptile-like bird Archaeopteryx, *from the lithographic stone of Solnhofen in Bavaria*[28]. *This first specimen was found in 1861, two years after Darwin's* Origin of Species *was published; although its interpretation was controversial, it provided the first fossil evidence of the possibility of evolutionary connections between separate animal classes.*

growing confidence in the general tendency of new discoveries. Pictet had expressed the situation perceptively, when he confessed that Darwin's *imagination* moved more swiftly than his own; for what was needed, in order to convert most palaeontologists to an evolutionary way of thinking, was not sudden illumination but a gradual stretching of the imagination. But where Darwin had only outlined the imaginative possibilities in an almost prophetic visionary style, the literally down-to-earth evidence of new fossil discoveries was progressively giving fresh scope for a more scientifically disciplined imagination. Increasingly it was seen that fossil species and genera were indeed found in temporal successions that could best be interpreted as truly evolutionary lineages. Furthermore, evolutionary theory actually had predictive power, since it could suggest (though not of course infallibly) what form hitherto undiscovered intermediates ('missing links') were likely to take.

The re-direction of palaeontology into evolutionary channels can be illustrated by the further development of research on the genealogy of the horse family. Gaudry had described the less specialised form *Hipparion*, but this was still very different from any earlier Tertiary mammals. In 1871, however, the young Russian palaeontologist Vladimir Kovalevsky (1842–1883) re-studied some of Cuvier's material in Paris and showed that *Anchitherium* provided a convincing intermediate linking *Hipparion* back to Cuvier's Eocene *Palaeotherium*[31]. Then the American palaeontologist O. C. Marsh (1831–1899), who like Kovalevsky had earlier come to study in Western Europe, began searching for evolutionary lineages in the rich fossil faunas of his own country. By 1874 he felt confident in announcing that the fossil record of the horse family in the New World was more complete than the European, and that he had an adequate series all the way from an unspecialised Eocene *Orohippus* to the highly specialised living *Equus*: "the line of descent appears to have been direct", he concluded, "and the remains now known supply every important intermediate form" (Fig. 5.6)[32]. Two years later Huxley, who was in the United States for a lecture tour, was sufficiently convinced to adopt Marsh's interpretation for his own lecture on the evolution of the horses, thereby accepting that the European fossils had merely been occasional immigrants from the New World. At the same time he predicted the probable form of an ancestral *Eohippus*, which was duly discovered only two months later.

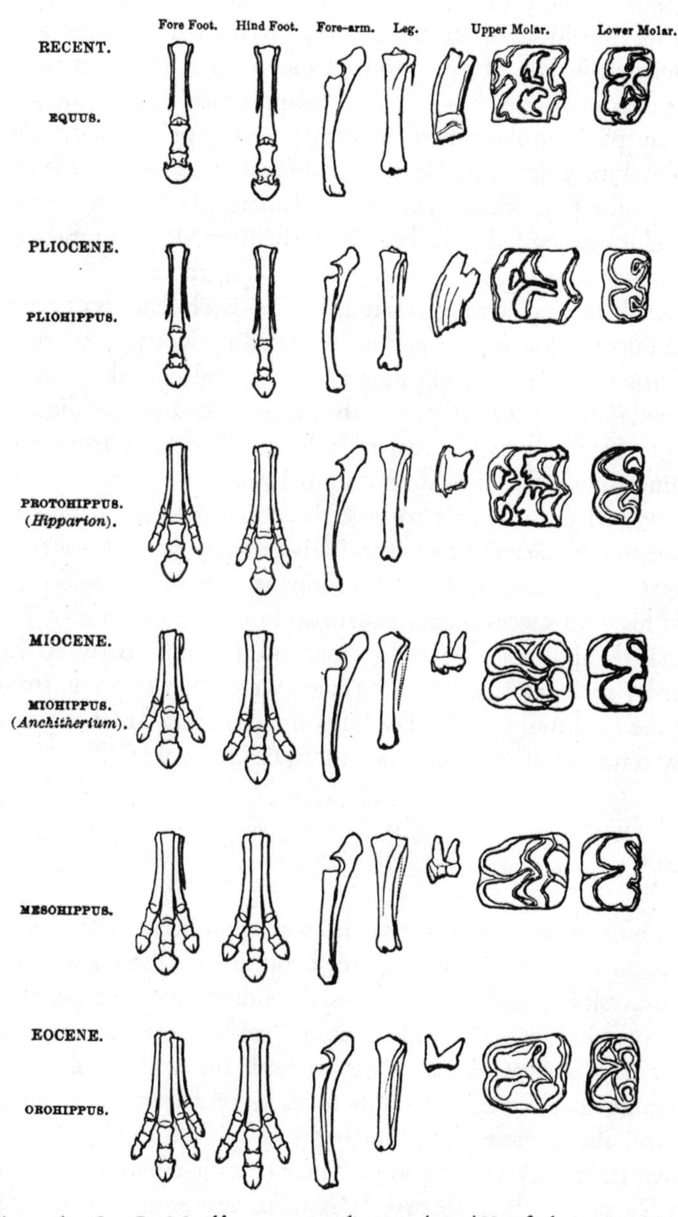

Fig. 5.6. O. C. Marsh's summary diagram (1879)[32] of the main, North American, line of evolution of the horses during Cainozoic time. In the 1870s this was one of the best-documented and most persuasive examples of evolution at the generic level.

Marsh's re-interpretation of the phylogeny of the horse family, relegating the older European discoveries to the status of offshoots of an American main line of evolution—a picture that has been confirmed by all subsequent research—symbolises appropriately the emergence of American palaeontology from its earlier quasi-colonial status into full intellectual maturity, The New World had long provided the research centres of the Old with abundant fossil materials, but on the intellectual level it had tended to be heavily derivative. Marsh is representative of the generation which, though still looking to Europe for some of their initial training, was to create major research schools of their own outside Europe. The same trend is illustrated by Kovalevsky, returning from Paris to begin a distinguished career in palaeontology in Russia. This trend is of course only part of the general expansion of the scientific enterprise at this period, outwards from Western Europe towards becoming fully international on a global scale.

By 1877 Marsh was able to give the American Association for the Advancement of Science an authoritative survey of the fossil record of American vertebrates in wholly evolutionary terms. He could point not only to his own successful reconstruction of the horse family, but also to Huxley's and his own indications of the links between reptiles and birds: such intermediates, he said, were "the stepping stones by which the evolutionist of today leads the doubting brother across the shallow remnant of the gulf, once thought impassable"[33].

IX

Marsh's metaphor of stepping-stones was appropriate, and it indicates the level on which the vindication of evolutionary theory was proving most successful. Fossil material was providing many convincing examples of phylogenetic series linking similar genera and families together, and at least a few hints of how the wider gulfs between classes might be crossed. The difficulties were still much greater both below and above these levels.

Below, there was the question of inter-specific evolution, the crucial step in Darwin's whole theory. Was there any positive evidence that species evolved into new forms by the extremely slow and gradual process that the hypothesis of natural selection implied? Darwin and Lyell, as we have seen, explained away the lack of fossil evidence of such transitions by appealing to the extremely fragmentary nature of the record; but most palaeontologists were unconvinced. One attempt was made to construct a phylogeny at the species level, using some highly diverse freshwater molluscs collected from ten successive zones of a Tertiary limestone[34]; but this example could be questioned on the grounds that the whole assemblage might constitute merely a single unusually variable population. It was not until 1875 that Melchior Neumayr (1845-1890), the professor of palaeontology at Vienna, published what was probably the first satisfactorily uninterrupted "series of forms" (*Formenreihe*) connecting fossil species. Neumayr subtitled his memoir explicitly "A contribution to the theory of descent". Arguing, as a convinced evolutionist, that it was impossible to conceive that species remain constant through geological time, he showed how certain Tertiary non-marine molluscs could be arranged to form evolutionary sequences[35]. Soon afterwards a brief paper was published in England describing the gradual changes in the form of the sea-urchin *Micraster* in successive strata of the Chalk, and pointing out that the directional nature of these changes was "suggestive rather of progressive development than of simple variation" (Fig. 5.7).

However such examples were very rare; and they received curiously little attention as crucial evidence for the gradual character of trans-specific evolution. The *Micraster* series, which even today is one of the best documented cases of micro-evolution in fossils, was virtually forgotten for twenty years before being re-studied independently and in greater detail—but ironically it was then presented primarily as an example of the precision with which fossils could be used stratigraphically, rather than as an example of evolution![36] The reason for this neglect of evidence for small-scale evolution is not altogether clear. Possibly palaeontologists were persuaded by the Darwinian argument into feeling that the deposition of strata must have been generally too discontinuous for there to be much hope of finding such evidence; yet the uniformity of certain formations (notably the Chalk) pointed to the possibility of fairly continuous deposition, which should have suggested that it was worth searching for evidence of slow evolution. A more likely reason is that by the 1870s, when the examples just mentioned were published, most palaeontologists had lost faith in the original Darwinian concept of extremely slow evolution by natural selection alone. They remained convinced that evolution had occurred, but not that its mechanism had been Darwinian.

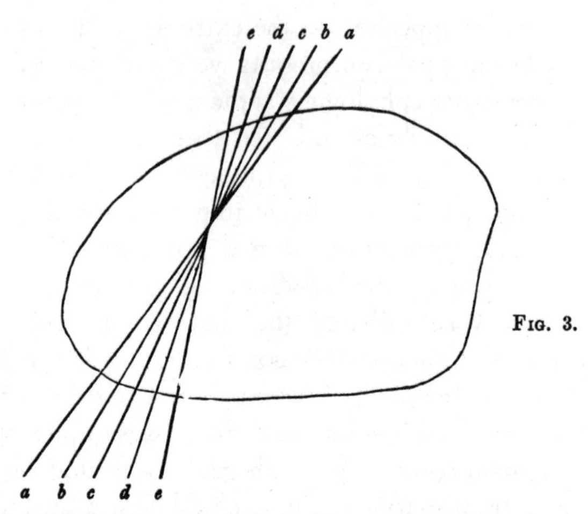

Fig. 3.—In which the changing relative positions of the mouth and apical disk are indicated by the lines *a-a*, *b-b*, etc.

Fig. 5.7. Meyer's diagram (1878) of the gradual directional *changes (a to e) in the positions of the mouth and apical disc of the fossil sea-urchin* Micraster, *from successive horizons in the English Chalk*[36]. *This was one of the few fossil examples of slow trans-specific evolution reported in the later nineteenth century, but the weight of other evidence* against *Darwinian evolution led to its neglect.*

The reasons for this rapid change of opinion, which affected even Darwin himself, can best be appreciated against the background of the problems set by the fossil record at the highest taxonomic level. We have seen that the search for fossil evidence for evolution was impressively successful at intermediate levels, and it became clear that many of the earlier 'laws' and generalisations of comparative anatomy, of embryology, of biogeography and even of palaeontology itself could be explained convincingly in evolutionary terms. But still there was a point at which the evidence from all these branches of biology ceased abruptly. The old concept of the 'unity of type', as modernised in Owen's theory of archetypes, could be re-interpreted as evidence for a common evolutionary origin, but only as far as the concept itself was valid. Cuvier had shown long before, and Owen had re-affirmed with a much greater mass of evidence, that unity of type was *not* valid above the level of the great *embranchements* or phyla of the animal kingdom. Comparative anatomy therefore offered no evidence for the common origin of all the major groups of animals. Von Baer had shown that the same limit applied in embryonic development. More recently, and most seriously, the unravelling of the Palaeozoic strata had showed that the major groups were clearly distinct in the oldest fossiliferous deposits.

A Darwinian imagination might overleap these barriers, and conceive that even such distinct kinds of organism might have been differentiated by slow evolution, but there was an embarassing lack of evidence for that conclusion. Huxley, as we have seen, extracted from the fossil record as much as he could in favour of evolution; but at this point even he was obliged to fall back on the imperfection of the fossil record, in a form only slightly less extreme than Darwin's and Lyell's. When reviewing the state of the evolutionary theory in 1870, in his Presidential Address to the Geological Society, Huxley had to postulate an extremely lengthy and almost unrecorded pre-Palaeozoic (that is, Pre-Cambrian) history of life. Even the earliest known (Silurian) fish, for example, were already highly complex vertebrates, far removed in organisation from what Huxley considered the original vertebrate stock must have been like: "it is appalling to speculate", he reflected, "upon the extent to which that origin must have preceded the epoch of the first recorded appearance of vertebrate life"[37]. Other scientists, however, were more appalled by the extent to which Huxley was prepared to rely on negative evidence.

It is only fair to add, however, that Huxley believed the pre-Palaeozoic strata would eventually yield positive evidence, and that in one instance they had already done so. He made full use of the discovery, in 1863, of some apparently organic structures in extremely ancient ('Laurentian') rocks in Canada. These structures had been hopefully named *Eozoon* (that is, dawn-animal), and were thought to be as much older than the Cambrian fossils as the latter were older than the faunas of the present day (Fig. 5.8). Within a few years, however, this evidence collapsed, with the demonstration that *Eozoon* was an inorganic structure of metamorphic origin; for the rest of the century the Pre-Cambrian was left devoid of any well authenticated fossils, and the (geologically) sudden appearance of the Palaeozoic organisms remained as mysterious as ever.

Meanwhile, however, Huxley's (and Darwin's) conjuring up of vast periods of Pre-Cambrian time, in order to accommodate the slow differentiation of the major groups of animals and plants, had come under very serious attack from a different direction. The distinguished physicist William Thomson (1824–1907), later Lord Kelvin, turned his attention in the 1860s from the problem of the origin of the Sun's

PLATE IV.

Magnified and Restored Section of a portion of Eozoon Canadense.

The portions in brown show the animal matter of the Chambers, Tubuli, Canals, and Pseudopodia; the portions uncoloured, the calcareous skeleton.

Fig. 5.8. Dawson's reconstruction of the Pre-Cambrian 'fossil' Eozoon *(1875)*[38]. *Although the organic origin of* Eozoon *was disputed and ultimately rejected, Dawson reconstructed it as a protozoan, and believed it extended the fossil record much further back in time than any fossil previously known. Others took this as evidence that there had been ample time for the animal phyla to have been differentiated by slow evolution.*

energy to the subsidiary problems of geophysics. This involved re-assessing Fourier's earlier calculations of the thermal history of the Earth in the light of the recently formulated laws of thermodynamics. Thomson was concerned above all that the speculations of geologists and biologists—the timing of his work strongly suggests that he had

Lyell and Darwin in mind—should not violate basic physical principles. Using the best available data on the geothermal gradient and the physical properties of rocks, and assuming that the Earth had cooled gradually from an originally molten state, Thomson calculated that about 98 million years (m.y.) had passed since the solidification of the crust. The figure had a spurious air of accuracy, since he admitted that the many assumptions necessary in the calculations allowed for a margin of error stretching from 20 to 400 m.y. But even the upper limit represented an extremely severe restriction on the Lyellian invocation of unlimited time as a explanatory device, since a substantial part of the 400 m.y. would need to be allowed for the period of cooling between the first consolidation of the Earth's crust and the reduction of the surface temperature to a level at which life would become possible[39].

Thomson's argument carried all the prestige of the most 'fundamental' of the sciences, and its mathematical rigour seemed impressive. It was a crushing blow to Lyell's steady-state interpretation of Earth-history, but its implications were even more disastrous for Darwin's biological theory. Darwin had rashly estimated no less than 300 m.y. for the erosion of the English Weald (that is, roughly for Cainozoic time alone)—an estimate which of course argued in favour of the efficacy of natural selection as the agency of extremely slow, and extremely poorly recorded, organic evolution. Darwin's prodigal use of a Lyellian time-scale had already been criticised by Phillips on purely geological grounds: using present rates of erosion and deposition as a guide (and thereby following the best Lyellian methods of actualism!) Phillips had estimated that the whole known sequence of strata might have taken about 96 m.y. to accumulate. Subsequent estimates by other geologists suggested that within the unavoidable margins of error the geological and the physical evidence were reasonably congruent. Whatever the exact figures suggested, the time scale could no longer be enlarged at will in order to save the phenomena for Darwinian evolution (it remained relatively constricted until the discovery of radioactivity upset Thomson's basic assumptions, but even today it has not expanded to the scale of Darwin's earlier estimates).

One of Thomson's close associates, the physicist and engineer Fleeming Jenkin, was the first to use Thomson's physics explicitly against the hypothesis of natural selection; and Jenkin's review of the *Origin* made the attack still more serious by giving also a mathematical demonstration that blending inheritance would make natural selection unworkable. Thomson then made his own attack more explicit, arguing in 1869 that unless biology was to disregard the basic physical 'laws of nature' it must accept a restricted time-scale within which evolution *by natural selection* was impossible. Thomson was not attacking evolution as such, but only its Darwinian mechanism.

Like many other critics of natural selection, Thomson had his own metaphysical motives: he believed that the general 'lawfulness' of Nature itself was at stake, since natural selection seemed to relegate the development of life ultimately to the realm of chance. But whatever motives may have been operative the fact remains that the hypothesis of natural selection was in full retreat within a dozen years of the first publication of the *Origin*, in the face of an impressive array of arguments from many different branches of science. Within a greatly restricted time-scale, on the validity of which most geologists agreed with the physicists, vast periods of unrecorded time could no longer be invoked so readily to explain each and every failure of the fossil record. No scientist was tempted to return to a pre-evolutionary interpretation of the fossil record; but a large question-mark had been put beside the Darwinian assumption that evolution must always be extremely slow and extremely gradual.

Darwin himself became increasingly worried by the problem, withdrew his earlier estimates of the time-scale, and gradually came to rely less and less on natural selection[40]. Wallace accepted the new turn of events more wholeheartedly, suggesting that about a quarter (24 m.y.) of Thomson's allowance could be assigned to the preserved fossil record (Cambrian to present), leaving the remainder available in the Pre-Cambrian for the differentiation of the major groups of organisms. But this necessarily involved postulating much more rapid rates of evolution in the past.

X

These developments virtually shelved the mechanism of natural selection for the rest of the century. But they had the beneficial effect of freeing evolutionary speculation from what could easily have become a mental straitjacket. In 1870, for example, Huxley felt that the only way to explain the sudden appearance of most of the orders of placental mammals at the beginning of the Cainozoic was to postulate an extremely long preceding period of very gradual (and unrecorded) differentiation. Once the assumption of uniformly slow evolution was questioned, however, the way was open to re-interpreting the fossil evidence in terms of a (geologically) rapid episode of 'adaptive radiation', in which the primitive placental stock could have evolved into a variety of ecological niches left vacant by the extinction of most of the Mesozoic reptiles.

Likewise the possibilities of relatively rapid or 'saltatory' trans-specific evolution were explored; and as already suggested this may account for the lack of research into cases of extremely gradual trans-specific change. For example, when in 1869 Wilhelm Waagen (1841–1900), a young German palaeontologist who was a keen evolutionist, re-interpreted a single 'species' of Jurassic ammonite as an evolutionary series, he did so in terms of successive "mutations"[41]. Waagen's' mutation theory' (which is distinct from the later genetical use of the same term) enjoyed much support among palaeontologists, for it accounted for the evolutionary links between successive fossil species without having to postulate unrecorded intermediate forms between them. It is possible that the very common use of 'chain' metaphors in the evolutionary terminology of this period (for example Gaudry's *liens* and *enchainements*) should be taken seriously as an indication that evolutionary change really was conceived generally as a series of small 'quantum' leaps from one well-adapted species to the next.

The proliferation of evolutionary speculations had, however, a less beneficial aspect. The failure of the fossil record to provide positive evidence for evolution at the higher taxonomic levels caused many evolutionists to fall back on the less direct evidence of comparative embryology. The remarkable 'metamorphoses' observed in the development of many organisms suggested an appropriate model for evolutionary change; and the old concept of individual development paralleling the Scale of Beings was rapidly refurbished in evolutionary form as a concept of 'ontogeny' recapitulating 'phylogeny'. This 'biogenetic law' was popularised, if not initiated, by the German biologist Ernst Haeckel, and had the baneful effect of deluding biologists into believing they possessed a virtually infalliable key to the evolutionary ancestry of living organisms. In the euphoric atmosphere produced by the consequent re-interpretation of morphology and embryology, the fossil evidence for evolution was often ignored or, at best, gratuitously moulded to suit pre-conceived conclusions.

Archetypes became hypothetical ancestors, with scant regard to their adaptive viability as living organisms; and the failure of palaeontologists to discover any such creatures in the fossil record was attributed, as usual, to its fragmentary character (Fig. 5.9).

Palaeontologists themselves did not remain immune from this fashion for evolutionary speculation. Paradoxically, this resulted in the loss, or at least submergence, of much that Darwin had stood for. The Cuvierian emphasis on the functional integrity of organisms, reinforced by Paleyan natural theology, had developed quite directly into the Darwinian emphasis on the centrality of adaptation and its causation[42]; but this functional emphasis faded almost completely from the study of fossils (and has only very recently begun to be recovered). Most palaeontologists forgot—at least in practice—that their fossils had once been living organisms, and that they had been adapted to some mode of life; and they tended to regard their specimens exclusively as evidence for evolutionary ancestry.

However this shift of emphasis was only part of a much wider trend. Emerging into scientific respectability in the 1860s, the theory of organic evolution was at once absorbed into the all-embracing philosophical system of the period. Even before the *Origin* was published the idea of gradual 'development' was being used as a unifying philosophical principle: it fitted neatly into the mid-century's confident optimistic belief in progress. The fashionable English philosopher of the age, Herbert Spencer, easily absorbed Darwin's evolutionary theory—or certain parts of it—into his system as merely the biological aspect of the much wider principle of 'development'. This absorption was not merely condoned but positively welcomed even by evolutionary biologists such as Huxley. When he introduced *Archaeopteryx* as a vivid piece of evidence for Darwin's theory, Huxley explicitly declared himself a firm believer in Spencer's doctrine of development, and explained that he regarded evolution as but a part of this all-embracing principle.

The same tendency to absorb Darwin's theory into a philosophical system can be seen even more clearly in Germany, where a cult of *Darwinismus* very quickly relegated Darwin's own thought to a minor position within a far wider programme of advocating a philosophy of monistic materialism. Darwin's theory was valued less for its potentially successful solution of a long-standing scientific puzzle, than as ammunition with which to blast the remnants of theism from the philosophical

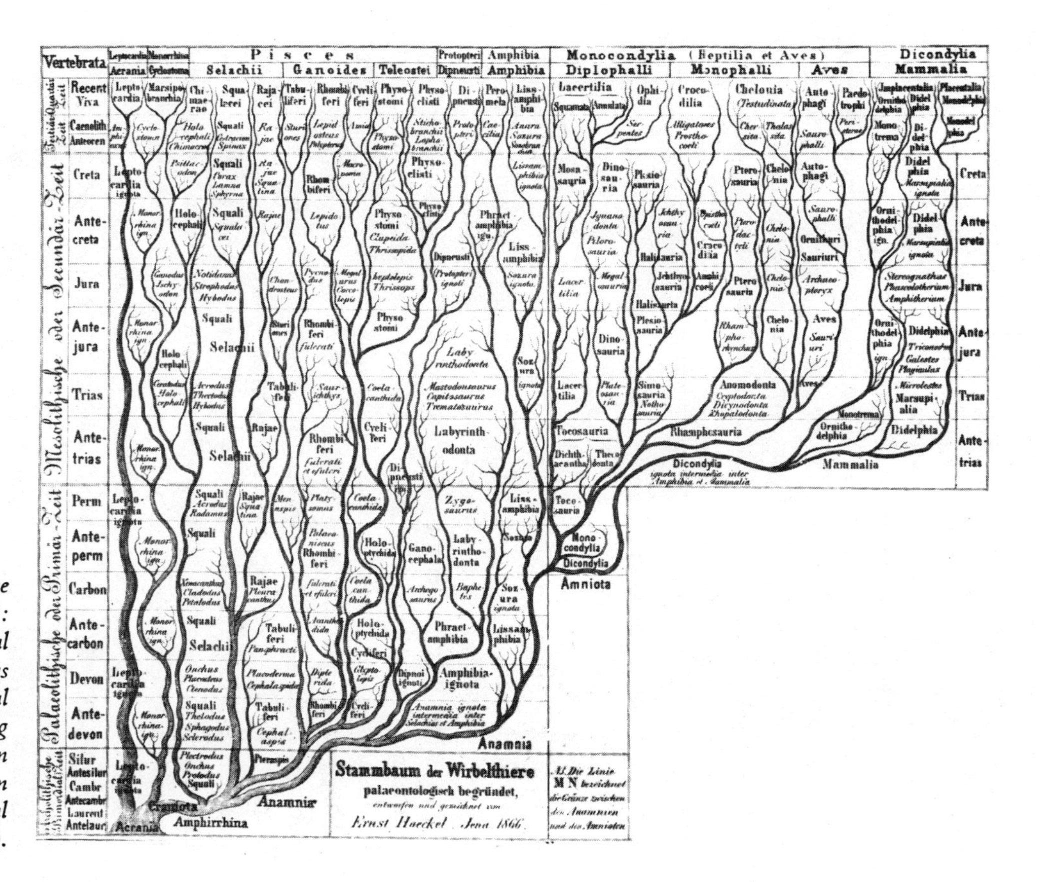

Fig. 5.9. *Ernst Haeckel's speculative 'phylogeny' of the vertebrates (1866)*[43]*: note how the traditional non-temporal 'tree' of hierarchical classification has been converted directly into a temporal evolutionary history. The embarrassing lack of fossil intermediates has been eliminated by the gratuitous assumption of large gaps in the stratigraphical record ('Ante-jura', 'Ante-trias', etc.).*

scene. This can be illustrated from Haeckel's popular exposition of evolutionary philosophy, the *Natural Creation-History* (1868), in which the emphasis was on the purely *natural* character of the developments conventionally termed 'creative'; but the subtitle placed Darwin in strange company—that of Lamarck and Goethe—as the most recent member of a pantheon of evolutionary philosophers[43]. In other words, Darwin's particular theory about evolution was less important to Haeckel than the fact that he had at last made an evolutionary theory *of any kind* "mechanistically based" and therefore scientifically respectable.

XI

To trace the development of palaeontology in the later nineteenth century, even in outline, would need much fuller treatment than is possible here. It is not inapprorpiate to bring these essays to a close in the 1870s, for despite the vast and exponential growth of palaeontology in the last hundred years its major features were already clear by that decade. Institutionally, it had spread outwards from Western Europe to all the 'developed' industrial nations, to become part of the world-wide scientific enterprise, marked by all the familiar features of research schools, societies, conferences, periodicals and international commuting. Its practical value to stratigraphy, and hence to the exploitation of mineral resources, had ensured it increasing governmental support through geological surveys; yet this benefit had its price, for the subservience of palaeontology to stratigraphy was already beginning to deflect its main sympathies away from the other biological sciences and to narrow its intellectual horizons—a trend that has only been reversed in very recent years. Conceptually, the general theory of evolution had provided it with a unifying principle of great explanatory value, enabling it to synthesise the ever-increasing bulk of detailed descriptions into a coherent picture of the history of life, however debatable the mechanisms of evolution remained; yet this too was achieved at the high price of neglecting the valuable insights of earlier traditions of interpretation, and—for many decades—of diverting theoretical energies into sterile speculations.

At the same time it is undeniable that palaeontology was withdrawing more and more from the position of intellectual importance that it had held in the public mind earlier in the century. In part this was no doubt an inevitable result of increasing specialisation and professionalisation, which removed the results of research further from the immediate comprehension of the interested layman, and also reduced his chances of making any substantial contribution to the science as an amateur. More fundamentally, however, the withdrawal of palaeontology from the public gaze was due to the failure of the fossil record to give much enlightenment on the questions that most agitated thinking men.

Earlier in the century the effect of palaeontology on this level had been profound, for it had disclosed at least the broad outlines of a history of life that had clear implications for human self-understanding. It had shown that that history had been of almost inconceivable length; that life on Earth had passed through many strange phases, becoming progressively more complex, more varied, and more like the living world of the present; that Man, although the most recent newcomer to the scene, could be interpreted as the supreme and crowning feature of the entire history; and above all, that this history was essentially intelligible and meaningful. Man might seem dwarfed by the magnitude of the geological eras that preceded him, but those eras could at least be seen as a long and patient preparation for his coming.

However, once attention turned to the *means* by which new kinds of life had come into existence, palaeontology proved unable to supply the insights that thinking men demanded. The imperfections of the fossil record were exploited systematically to undermine even the expectation of discovering such insights; but above all the fossil record had too little light to throw on the origin and nature of Man himself. For the concern with Man's place in nature was far more central, even in the minds of those we label 'scientists', than their published technical works may suggest; and metaphysical concerns with the meaning of the natural world were more pressing than an older and positivistic history of science allowed. But with the possible exception of studies of human and primate fossils, such concerns received relatively little clear satisfaction from the findings of palaeontology in the later nineteenth century, and public interest in the science declined accordingly. Perhaps as a result, it would seem that palaeontology—and indeed geology too —failed to gain a proportionate number of recruits of first-class intellectual calibre. The output of specialist papers and monographs continued to rise exponentially, but their character became routine and their intellectual level stagnant. Only recently have there been hopeful signs that palaeontology may be recovering, in its younger generation, the broad interests and outlook that it possessed so markedly earlier in its history.

XII

As palaeontology now prepares, therefore, for a great leap forward into a computerised age (for which the nature of its material makes it highly appropriate) there is perhaps a danger that it may lose sight of its historic origins in the 'steam age' of science and before. That it should not become a-historical in outlook is important, not for nostalgic antiquarian reasons, but because the loss of historical perspective would lead to conceptual impoverishment. In every period of its history, palaeontology, like all other branches of science, has developed through a series of intricate interactions between philosophical presuppositions (often implicit or even unrecognised), theoretical constructions at all levels, and the steadily accumulating fund of observational evidence. An exclusive pre-occupation with the last of these, however excusable it may be in the heat of the present information explosion, will not lead to a more securely and factually based science, but quite probably to a vast superstructure built on unexamined and perhaps weak conceptual foundations. Reflection on the history of palaeontology, with its reminder of the very different worlds of thought in which the science acquired the various strands of its present complex texture, may perhaps help in the critical examination and re-appraisal of its present foundations, and thus ensure that a computerised access to its vast stores of factual information is used to the best heuristic advantage.

For the historian of science, it is perhaps useful to reflect on the complex relation between continuity and discontinuity in the conceptual development of palaeontology. Even more than in other branches of science, the element of cumulative continuity has visible and tangible expression in the steadily growing bulk of fossil material preserved in museums and research institutions. Yet it is no longer adequate to regard this as parallel in any simple way to the conceptual growth of the subject. The 'meaning' of fossils has been seen in many different ways in different periods. Indeed, the same fossil specimens (for example sharks' teeth) have been re-interpreted several times within different frames of reference—they have, as it were, been seen with different eyes. However there is now a danger that the currently fashionable emphasis on the discontinuities of interpretation may be taken too far. It would be salutary for the historians to be reminded by the palaeontologists that most of the earlier frames of reference are still being utilised—even if unrecognised—in modern palaeontology: the insights and methods of one 'paradigm' of interpretation have not been wholly abandoned, but absorbed into the next. The debate on the origin of fossils, which was central to the future science of palaeontology in the age of Gesner, is still alive whenever questions of fossilisation or the origins of Problematica are being discussed. The idea of using fossils to construct a chronology of Earth-history, which was central in the age of Steno and Hooke, and which drew its inspiration from contemporary historical scholarship, is still alive whenever palaeontologists tackle the problems of stratigraphical correlation. The concept of the 'designful' adaptive integrity of extinct species, which was central in the age of Cuvier and Buckland, is still alive whenever the functional morphology and ecology of fossil organisms is being reconstructed. The debate about the gradual or paroxysmal, uniform or fluctuating, rate of organic change, which was central in the age of Lyell and Murchison, is still alive wherever the problems of 'tempo and mode' in evolution and extinction are under discussion. Finally, the debate on the exact nature of species production, which was central in the age of Bronn and Darwin, is still alive whenever palaeontology contributes to the continuing research on the mechanisms of evolutionary change.

REFERENCES

1. *Comptes-Rendus hebdomadaires de l'Académie des Sciences*, vol. 44, pp. 166–7 (1857). Bronn's essay was first published as *Untersuchungen über die Entwickelungs-Gesetze der organischen Welt während der Bildungs-Zeit unsere Erd-Oberfläche*, Stuttgart, 1858; and afterwards by the Academy as 'Essai d'une réponse à la question de prix proposée en 1850 . . . ', *Supplément aux Comptes-Rendus des Séances de l'Academie des Sciences*, vol. 2, pp. 377–918 (1861).

2. *Comptes-Rendus*, vol. 30, pp. 257–260 (1850).

3. See Walter F. Cannon, 'The Uniformitarian-Catastrophist Debate', *Isis*, vol. 51, pp. 38–55 (1960).

4. Heinrich Georg Bronn, *Italiens Tertiär-Gebilde und deren organische Einschlusse, Vier Abhandlungen*, Heidelberg, 1831; *Lethaea Geognostica, oder Abbildungen und Beschreibungen der für die Gebirge-Formationen bezeichnendsten Versteinerungen*,

Stuttgart, 1835–8; *Handbuch der Geschichte der Natur*, Stuttgart, 1841–2; *Index Palaeontologicus oder Übersicht der bis jetzt bekannten fossilen Organismen*, Stuttgart, 1848–9.

5. F. Unger, *Versuch einer Geschichte der Pflanzenwelt*, Wien, 1852.

6. See also H. G. Bronn, *Morphologische Studien über die Gestaltungs-Gesetze der Naturkörper überhaupt und der organischen inbesondere*, Leipzig und Heidelberg, 1858.

7. See G. Canguilhem *et al.*, 'Du développement à l'évolution au XIXe siècle', *Thalès*, vol. 11, pp. 1–68 (1962); Jane Oppenheimer, 'An embryological enigma in the Origin of Species', *in* Bentley Glass (ed.), *Forerunners of Darwin: 1745–1859*, Baltimore, 1959: pp. 292–322.

8. Karl Ernst con Baer, *Über Entwickelungsgeschichte der Thiere. Beobachtung und Reflexion*, Königsberg, 1828, 1837; see also Elizabeth B. Gasking, *Investigations into Generation 1651–1828*, London, 1967.

9. Alfred R. Wallace, 'On the Law which has regulated the Introduction of New Species', *Annals and Magazine of Natural History*, series 2, vol. 16, pp. 184–196 (1855); reprinted in C. F. A. Pantin, 'Alfred Russel Wallace: his pre-Darwinian essay of 1855', *Proceedings of the Linnean Society of London*, vol. 171, pp. 139–153 (1960); Charles Darwin and Alfred Wallace, 'On the Tendency of Species to form Varieties: and on the Perpetuation of Varieties and Species by Natural Means of Selection', *Journal of the Linnean Society of London* (*Zoology*), vol. 3, pp. 45–62 (1859), reprinted in *Evolution by Natural Selection* (ed. Gavin de Beer), Cambridge, 1958, pp. 255–279.

10. Charles Lyell, 'Anniversary Address[es] of the President', *Quarterly Journal of the Geological Society of London*, vol. 6, pp. xxvii–lxvi (1850); vol. 7, pp. xxv–lxxvi (1851).

11. G. de Beer, 'Darwin's notebooks on the transmutation of species', *Bulletin of the British Museum* (*Natural History*), *Historical series*, vol. 2, parts 2–6 (1960–61), vol. 3, part 5 (1967); S. Smith, 'The origin of 'The Origin' as discerned from Charles Darwin's notebooks and his annotations in the books he read between 1837 and 1842', *Advancement of Science*, vol. 16, pp. 391–401 (1960).

12. Robert M. Young, 'Malthus and the Evolutionists: the common context of biological and social theory', *Past and Present*, no. 43, pp. 109–145 (1969).

13. The 1842 Sketch is printed in Darwin and Wallace, *Evolution by Natural Selection* (ed. Gavin de Beer), Cambridge, 1958, pp. 39–88: the palaeontological section is at pp. 59–63.

14. The 1844 essay is printed in Darwin and Wallace (see note 13), pp. 89–254: the palaeontological section is Chapter 4, pp. 154–162.

15. Charles Darwin, 'A Monograph on the Fossil Lepadidae, or Pedunculated Cirripedes of Great Britain', *Monographs, Palaeontographical Society*, 1851; 'A Monograph on the Fossil Balanidae and Verrucidae of Great Britain', *ibid.*, 1854.

16. See Walter F. Cannon, 'Darwin's Vision in *On the Origin of Species*', *in* G. Levine and W. Madden (eds.), *The Art of Victorian Prose*, New York, 1968, pp. 154–176; Robert M. Young, 'Darwin's Metaphor: does Nature select?,' *The Monist*, vol. 55, pp. 442–503 (1971).

17. Charles Darwin, *On the Origin of Species by means of Natural Selection, or the Preservation of favoured Races in the Struggle for Life*, London, 1859; facsimile reprint (ed. Ernst Mayr), Cambridge (Mass.), 1964; also reprinted (ed. J. W. Burrow), London, 1968. See also Gavin de Beer, *Charles Darwin. Evolution by Natural Selection*, London, 1963.

18. F.-J. Pictet, 'Sur l'Origine de l'Espèce, par Charles Darwin', *Bibliothèque universelle. Revue suisse et étrangère*, (n.pér.), vol. 7, *Archives des Sciences physiques et naturelles*, pp. 233–255 (1860).

19. F.-J. Pictet, *Traité de Paléontologie, ou Histoire naturelle des animaux fossiles considerés dans leurs rapports zoologiques et géologiques*, Paris, 1844–46.

20. John Phillips, *Life on the Earth: its Origin and Succession*, Cambridge and London, 1860.

21. See A. Ellegard, *Darwin and the general reader. The reception of Darwin's theory of evolution in the British periodical press, 1859–1872*, Göteberg, 1958; Robert M. Young, 'The impact of Darwin on conventional thought', *in:* Anthony Symondson (ed.), *The Victorian Crisis of Faith*, London, 1970, pp. 13–35.

22. [Owen], *Edinburgh Review*, vol. 111, pp. 487–532 (1860); see Roy M. MacLeod, 'Evolutionism and Richard Owen 1830–1868: an episode in Darwin's century', *Isis*, vol. 56, pp. 259–280 (1965).

23. Richard Owen, *Palaeontology, or a systematic summary of extinct animals and their relations*, London, 1860.

24. Leonard G. Wilson, *Sir Charles Lyell's Scientific Journals on the Species Question*, New Haven, 1970.

25. Charles Lyell, *The geological evidences of the antiquity of man, with remarks on the origin of species by variation*, London, 1863; *Principles of Geology, or the modern Changes of the Earth and its Inhabitants*, 10th edition, 2 vols, London, 1868.

26. See K. P. Oakley, 'The problem of Man's antiquity', *Bulletin of the British Museum (Natural History), Geological series*, vol. 9, no. 5 (1964); J. W. Gruber, 'Brixham Cave and the antiquity of man', *in* Melford E. Spiro (ed.), *Context and Meaning in Cultural Anthropology*, New York, 1965, pp. 373–402; Boucher de Perthes, *Antiquités celtiques et antediluviennes. Mémoire sur l'Industrie primitive et les arts à leur origine*, 3 vols, Paris, 1847–64.

27. T. H. Huxley, *Evidence as to Man's Place in Nature*, London, 1863 (reprinted in *Collected Essays*, vol. 7, pp. 1–208, 1895).

28. Albert Gaudry, *Animaux fossiles et géologie de l'Attique d'après les recherches faites en 1855–56 et 1860 sous les auspices de l'Académie des Sciences*, Paris, 1862–67.

29. Owen, 'On the *Archaeopteryx* of von Meyer, with a description of the Fossil Remains of a Long-tailed species, from the Lithographic Stone of Solenhofen', *Philosophical Transactions of the Royal Society of London*, vol. 153, pp. 33–47 (1863); S. J. Mackie, 'The Aeronauts of the Solenhofen Age', *The Geologist*, vol. 6, pp. 1–8 (1863). See also Gavin De Beer, *Archaeopteryx lithographica. A study based on the British Museum specimen*, London, 1954.

30. Huxley, 'On the animals which are most nearly intermediate between birds and reptiles', *Annals and Magazine of natural History*, ser. 4, vol. 2, pp. 66–75 (1868); 'Further evidence of the Affinity between the Dinosaurian Reptiles and Birds', *Quarterly Journal of the Geological Society of London*, vol. 26, pp. 12–31 (1870); A. Wagner, 'Neue Beiträge zur Kenntnis der urweltlichen Fauna des lithographischen Schiefers [Part 2]', *Abhandlungen der königlichen bayerischen Akademie der Wissenschaften*, Klasse 2, Band 9, Abtheilung 1 (1861).

31. W. Kovalevsky, 'Sur l'Anchitherium aurelianense Cuv. et sur l'histoire paléontologique des Chevaux', *Mémoires de l'Academie imperiale des Sciences de St Petersbourg*, série 7, vol. 20, no. 5 (1873). See also A. Borissiak, 'W. Kowalewsky, sein Leben und sein Werk', *Palaeobiologie*, vol. 3, pp. 131–256 (1930).

32. O. C. Marsh, 'Notice of new equine mammals from the Tertiary formation', *American Journal of Science*, series 3, vol. 7, pp. 247–258 (1874); 'Polydactyl Horses, recent and extinct', *ibid.*, vol. 17, pp. 499–505 (1879). See also Charles Schuchert and Clara Mae Le Vene, *O. C. Marsh, Pioneer in Paleontology*, New Haven, 1940, ch. 9; G. G. Simpson, *Horses. The Story of the Horse Family in the Modern World and through Sixty Million Years of History*, New York, 1951, ch. 10.

33. O. C. Marsh, 'Introduction and Succession of Vertebrate Life in North America', *Nature*, vol. 16, pp. 448–450, 470–2, 489–491 (1877); 'Odontornithes: a Monograph of the Extinct Toothed Birds of North America', *Memoirs of the Peabody Museum, Yale University*, vol. 1 (1880).

34. Hilgendorf, 'Ueber Planorbis multiformis im Steinheimer Süsswasserkalk', *Monatsberichte der königlichen preussischen Akademie der Wissenschaften der Berlin*, vol. for 1866, pp. 474–504 (1866).

35. M. Neumayr and C. M. Paul, 'Die Congerien- und Paludinen-Schichten Slavoniens und deren Faunen. Ein Beitrag zur Descendenz-Theorie', *Abhandlungen der königlichen geologischen Reichsanstalt, Wien*, vol. 7, heft 3 (1875).

36. C. J. A. Meyer, 'Micrasters in the English Chalk—Two or more species?', *Geological Magazine*, new series, vol. 5, pp. 115–117 (1878); A. W. Rowe, 'An Analysis of the Genus *Micraster*, as determined by rigid zonal collection from the Zone of *Rhynchonella Cuvieri* to that of *Micraster coranguinum*', *Quarterly Journal of the Geological Society of London*, vol. 55, pp. 494–546 (1899).

37. T. H. Huxley, 'Anniversary Address of the President', *Quarterly Journal of the geological Society of London*, vol. 26, pp. xxix–lxiv (1870); reprinted in *Collected Essays*, vol. 8, pp. 340–388 (1894).

38. J. W. Dawson, 'On the Structure of certain Organic Remains in the Laurentian Limestone of Canada', *Quarterly Journal of the geological Society of London*, vol. 21, pp. 51–59 (1865); *Life's Dawn on Earth, being the History of the oldest known fossil Remains, and their Relations to Geological Time and to the Development of the Animal Kingdom*, London, 1875; Charles O'Brien, '*Eozoön Canadense*, "The Dawn Animal of Canada" ', *Isis*, vol. 61, pp. 206–223 (1970).

39. Joe D. Burchfield, 'Darwin and the dilemma of geological time', *Isis*, vol. 65, pp. 300–321 (1974); *Lord Kelvin and the Age of the Earth*, New York, 1975.

40. F. Jenkin, 'Darwin and the Origin of Species', *North British Review*, vol. 46, pp. 277–318 (1867). See P. Vorzimmer, 'Darwin and Blending Inheritance', *Isis*, vol. 54, pp. 371–390 (1963); B. G. Beddall, 'Wallace, Darwin and the theory of Natural Selection', *Journal of the History of Biology*, vol. 1, pp. 261–323 (1968); G. L. Geison, 'Darwin and Heredity: the evolution of his hypothesis of Pangenesis', *Journal of the History of Medicine*, vol. 24, pp. 375–411 (1969).

41. W. Waagen, 'Die Formenreihe des Ammonites subradiatus. Versuch einer paläontologischen Monographie', *Geognostische und Paläontologische Beiträge*, series 2, vol. 2, pp. 181–256 (1869).

42. See Walter F. Cannon, 'The bases of Darwin's achievement—a revaluation', *Victorian Studies*, vol. 5, pp. 109–134 (1961).

43. Ernst Haeckel, *Naturliche Schöpfungs-Geschichte. Gemeinverständliche wissenschaftliche Vorträge über die Entwickelungs-Lehre im Allgemeinen und diejenige von Darwin, Goethe und Lamarck im Besonderen*, Berlin, 1868; *Generelle Morphologie der Organismen. Allgemeine Grundzüge der organischen Formen-Wissenschaft, mechanisch begründet durch die von Charles Darwin reformirte Descendenz-Theorie*, Berlin, 1866.

Glossary

Adit	Mine-shaft driven horizontally.
Ammonite	Member of an order of extinct cephalopod molluscs, with a shell coiled in a plane spiral and divided into numerous chambers of complex form; distantly related to the living 'pearly nautilus.'
Arthropod	Member of a phylum of invertebrate animals, with an external skeleton and jointed limbs: examples include lobsters, spiders and insects.
Belemnite	Member of an order of extinct cephalopod molluscs, with a solid bullet-shaped 'guard' projecting from a conical shell divided into chambers; distantly related to the living cuttle-fish.
Bivalve	Member of a large class of molluscs, with a shell composed of two pieces hinged together; examples include cockles and mussels.
Boulder-clay	A distinctive deposit formed beneath ice-sheets in glaciated regions, consisting of clay containing angular stones and boulders of mixed sizes; widespread as a 'superficial' deposit in regions affected by the Pleistocene 'Ice Age.'
Brachiopod	Member of a phylum of invertebrate animals ('lamp-shells') with a shell composed of two pieces hinged together, having some resemblance to bivalves although unrelated; abundant as fossils, rare at the present day.
Cainozoic	Most recent of the major eras of geological time; on modern radiometric dating, from about 70 million years ago, to the present.

Calcareous Algae	Simple aquatic plants, related to the ordinary sea-weeds of the sea-shore, but secreting a skeleton of calcium carbonate, which can be preserved in the fossil state.
Calcite	Mineral composed of calcium carbonate; the commonest mineral of which the shells of 'shell-fish' (for example molluscs and brachiopods) are composed.
Cambrian	Earliest of the periods into which the Palaeozoic era is divided; the earliest strata with tolerably common fossils.
Carboniferous	One of the periods into which the Palaeozoic era is divided; the period during which most of the major coal deposits were formed.
Cast	Fossil in which the original material, for example of a shell or bone, has been replaced by a different mineral; also, the impression left by, for example, a shell on the consolidated sediment surrounding it.
Cephalopod	Member of a large class of marine molluscs; most living cephalopods are active swimmers, for example cuttle-fish, squid, octopus; fossils include ammonites and belemnites.
Cidaroid	Member of a distinctive order of echinoids (sea-urchins).
Class	In taxonomy, a major subdivision of a phylum; examples include Mammalia, Reptilia.
Cleavage	Tendency of crystals of certain minerals to break easily along planes oriented in particular directions.
Coal Measures	Formation of Carboniferous age in Europe and North America; includes most of the economically important coal seams of the coal-fields of the northern hemisphere.
Coelacanth	Member of a distinctive group of fish, thought to have been extinct since Cretaceous time, until a living species was discovered in 1939.
Concretion	Mineral mass, often of distinctive shape, enclosed in a sediment of different material.
Correlation	In stratigraphy, the determination of equivalent strata or formations in separate areas, often with the help of characteristic fossils.

Cretaceous	Youngest of three periods into which the Mesozoic era is divided; includes the Chalk as an especially distinctive formation.
Crinoid	Member of a class ('sea-lilies') of echinoderms, with superficially plant-like form; common as fossils; all but a few living species have lost the stem or 'stalk' and swim freely in the sea.
Crustacean	Member of a large class of aquatic arthropods; includes lobsters, crabs and shrimps.
Crystallography	Science of form and structure of crystals.
Dendritic markings	Superficially fern-like markings on surfaces of some rocks, formed inorganically by the crystallisation of mineral matter.
Echinoid	Member of a class ('sea-urchins') of echinoderms, with a roughly globular 'shell' (in fact an internal skeleton) usually covered with movable spines.
Eocene	One of the periods into which the Cainozoic era is divided.
Family	In taxonomy, a group of related genera.
Formation	Distinctive series of strata, for example, Chalk, Coal Measures.
Gastropod	Member of a large class of molluscs, generally with a coiled shell not divided into chambers; examples include snails and whelks.
Genus	In taxonomy, a group of related species (plural: genera).
Gneiss	Metamorphic rock of distinctive appearance, coarsely banded with different crystalline minerals; similar in composition to granite.
Guard	Solid bullet-shaped part of a belemnite, composed of calcite crystals radiating outwards.
Homology	In comparative anatomy, the relationship between equivalent parts in different organisms; for example, between the wing of a bat and the arm of a man.
Hypertropical	Climate even hotter than the tropical climate of the world at the present day.

Igneous	Process involving high temperatures, and some melting, within the Earth; rocks formed by such processes, for example, lava.
Jurassic	Middle of the three periods into which the Mesozoic era is divided; named after the Jura mountains in Switzerland.
Kupferschiefe	Formation of Permian age in Germany, with some strata with a relatively high content of copper.
Lias	Formation of Jurassic age in Western Europe, mostly shales and thin limestones, named from an English quarrymen's term.
Lithology	General character of a rock, for example, texture, nature of constituent minerals, etc.
Marcasite	Mineral composed of iron sulphide, often found as nodules (for example, in chalk) with crystals radiating from the centre.
Marker-horizon	Stratum or formation retaining especially distinctive lithology or fossils over a wide area, valuable for correlation.
Marsupial	Member of a group of mammals in which the young are born at an early stage of development and are then transferred to a pouch in the mother, for example, kangaroos.
Mesozoic	Era of geological time, with life of intermediate character between the Palaeozoic and the Cainozoic; on modern radiometric dating, from about 230 to about 70 million years ago.
Metamorphism	Series of processes, involving high temperature or pressure or both, resulting in major changes in the mineral texture or composition of a rock; examples include transformation of clay or shale into slate or schist, limestone into marble.
Miocene	One of the periods into which the Cainozoic era is divided.
Mollusc	Member of a very large phylum of invertebrate animals, including bivalves, gastropods and cephalopods; that is, most common 'shell-fish'.
Morphology	Science of form in organisms; also, the general structure and anatomy of an organism.

Muschelkalk	Distinctive limestone formation of Triassic age in Germany.
Nautilus	Rare surviving member of an order of cephalopod molluscs, of which extinct fossil forms are abundant; has a large spiral chambered shell with pearly interior.
Neopilina	Rare marine mollusc, belonging to archaic class that was thought to have been extinct since the Palaeozoic, until a living species was discovered in the 1950s.
Nova	'New' star; one that suddenly increases in apparent brightness.
Old Red Sandstone	Formation of Devonian age in Britain and elsewhere, formed under non-marine conditions and containing few fossils except early fish.
Oolite	Formation of Jurassic age in England, in which limestones of a distinctive texture ('oolitic') are prominent.
Order	In taxonomy, a division subordinate to a class, but superior to a family, for example the mammalian order Carnivora (adjective: ordinal).
Ossicle	In a crinoid, a segment of the jointed stem.
Osteology	Anatomical study of the bones of vertebrates; the structure of the skeleton of a particular vertebrate.
Oviparous	Reproducing by means of eggs laid externally, for example reptiles and birds.
Pachyderm	Member of an order of mammals in Cuvier's classification, including elephants.
Palaeolithic	'Old Stone Age,' characterised by chipped flint tools.
Palaeozoic	Earliest of three major eras of geological time with good fossil record; on modern radiometric dating, from about 570 to about 230 million years ago.
Permafrost	Permanently frozen ground in arctic and subarctic regions.
Permian	Youngest of the periods into which the Palaeozoic era is divided.

Phylum	In taxonomy, one of the major divisions of the animal kingdom, for example, Mollusca, Chordata (that is, all the vertebrates plus a few related animals), Arthropoda.
Placental	Mammals in which the young are retained in the mother's womb until a late stage of development; that is, almost all mammals except the marsupials.
Pleistocene	Geologically recent division of the Cainozoic era, roughly covering the duration of the 'Ice Age', that is from 1–2 million years ago until about 10,000 years ago.
Pliocene	Period of the Cainozoic era, preceding the Pleistocene.
Pre-Cambrian	Geological time before the beginning of the Palaeozoic era and Cambrian period, that is before about 570 million years ago; rocks with extremely rare fossils.
Problematica	Fossils of uncertain biological affinities or, in some cases, uncertain organic origin.
Pyrite	Mineral composed of iron sulphide, often occurring in shiny brass-coloured crystals ('fool's gold').
Quartz	Extremely common mineral composed of silicon oxide; usual constituent of common sand; gem quality is 'rock crystal.'
Radiometric dating	Method of dating rocks by measuring products of radioactive decay of certain elements; in modern geology, gives ages in millions of years (with substantial margins of error), supplementing but not supplanting older methods of 'relative' dating based on fossils.
Schist	Metamorphic rock, commonly breaking easily into sheets rather like slate.
Shale	Sediment of clay-like composition, compacted into easily-splitting layered structure.
Spar	Mineral with lustrous crystalline texture, for example calcite, fluorspar.
Speciation	Process by which new species are evolved from pre-existing ones.

Stalactite — Icicle-like mass of calcium carbonate hanging from roofs of limestone caves, formed by evaporation of dripping water.

Stalagmite — Pinnacle or layer of calcium carbonate formed on floors of limestone caves.

Stratigraphy — Study of stratified rocks, and their correlation between different areas.

Stratum — Layer of sedimentary rock, for example of clay, sandstone, limestone (plural: strata).

Superficial deposits — Deposits irregularly overlying a series of uniform strata or other rocks; examples include river gravels and boulder-clay.

Taxonomy — Study of classification of organisms; their naming and hierarchical arrangement, for example in phyla, classes, orders, families, genera and species.

Tectonic — Relating to the structure of the Earth's crust, and to the processes affecting its structure, for example mountain-building movements.

Tertiary — Old term for strata more recent than the 'Secondary' or Mesozoic; now almost synonymous with Cainozoic (but excluding 'Quaternary', that is Pleistocene and 'Recent').

Tetrapod — Vertebrate with four limbs, that is mammals, birds, reptiles and amphibians.

Triassic — Earliest of three periods into which the Mesozoic era is divided.

Trilobite — Member of a large class of extinct marine arthropods, with segmented external skeleton and prominent 'compound' (many-lensed) eyes.

Tuberculate — In an echinoid (sea-urchin), 'shell' covered with knobs bearing movable spines.

Ungulate — Member of order of mammals, with hoofed toes; examples include pigs, deer, sheep, cattle ('even-toed'), and horse and rhinoceros ('odd-toed').

Viviparous — Reproducing with young born 'alive' (that is, not enclosed in egg), for example almost all mammals.

Wenlock Limestone	Distinctive formation of Silurian age in England, forming prominent line of hills (Wenlock Edge).

Further Reading

Karl Alfred von Zittel, *Geschichte der Geologie und Paläontologie bis Ende des 19. Jahrhunderts*, Munich und Leipzig, 1899. (The English translation, *History of Geology and Palaeontology*, trans. M. M. Ogilvie-Gordon, London, 1901 (reprinted Weinheim 1962), unfortunately lacks all the references to the works cited, and is substantially abridged.) An exhaustive compilation, valuable as a source of references and factually reliable, but with little historical interpretation.

Archibald Geikie, *The Founders of Geology*, London, 1897 (second edition, London, 1905, reprinted New York, 1962). Although old-fashioned in approach, still a readable introduction to the subject.

Frank Dawson Adams, *The Birth and Development of the Geological Sciences*, London, 1938 (reprinted New York, 1954). Anecdotal and antiquarian in approach, but contains useful references and quotations, especially from works of the earlier periods.

Kirtley F. Mather and Shirley L. Mason, *A Source Book in Geology*, New York, 1939 (reprinted Cambridge (Mass.), 1970). An anthology of extracts, generally very short, from the works of geologists and palaeontologists up to the end of the 19th century.

Charles Coulston Gillispie, *Genesis and Geology. A study in the relations of scientific thought, natural theology, and social opinion in Great Britain, 1790–1850.* Cambridge (Mass.), 1951 (reprinted 1969). An important study arguing for the centrality of 'providentialism' in early 19th-century *British* geology.

Carl Chr. Beringer, *Geschichte der Geologie und des Geologisches Weltbildes*, Stuttgart, 1954. The best short interpretative account of the history of the earth sciences; particularly valuable for its balanced assessment of the work of geologists of different nationalities.

Loren Eisley, *Darwin's Century. Evolution and the Men who Discovered It*, London, 1959. A fairly reliable introductory account of evolutionary biology in the 19th century.

John C. Greene, *The Death of Adam. Evolution and its Impact on Western Thought*, Ames (Iowa), 1959. An attractively written and illustrated account of the history of evolutionary theory, particularly as it related to the place of Man in nature.

Francis C. Haber, *The Age of the World. Moses to Darwin*, Baltimore, 1959. A useful study, mainly on the 17th to 19th centuries, including much on the interpretation of fossils.

Helmut Hölder, *Geologie und Paläontologie in Texten und ihrer Geschichte*, Freiburg/München, 1960. A massive compilation, arranged by topics, with extensive quotations.

Stephen Toulmin and June Goodfield, *The Discovery of Time*, London, 1965. A wide-ranging and provocative essay on the history of the idea of nature's history.

W. N. Edwards, *The Early History of Palaeontology*, London, 1967 (first published 1931). A short and mainly derivative treatment, but with some interesting illustrations.

Cecil J. Schneer (ed.), *Toward a History of Geology*, Cambridge (Mass.), 1969. A collection of specialist papers by historians of geology and by geologists.

R. Hooykaas, *Continuité et Discontinuité en Géologie et Biologie*, Paris, 1970 (revised edition of *Natural Law and Divine Miracle. A historical-critical study of the principle of uniformity in geology, biology and theology*. Leiden, 1959). An important study of the meanings of 'uniformity' in the 'historical' sciences, with many examples from 19th-century geology and palaeontology.

Index